# Practical Handbook of Thematic Cartography

# Practical Handbook of Thematic Cartography
## Principles, Methods, and Applications

Nicolas Lambert and Christine Zanin

CRC Press
Taylor & Francis Group
Boca Raton London New York

CRC Press is an imprint of the
Taylor & Francis Group, an **informa** business

DUNOD Editeur, S.A, Foreign Language Publishing agreement
Manuel de cartographie. Principes, methodes, applications
By Nicolas LAMBERT and Christine ZANIN
Cartography by Nicolas LAMBERT and Christine ZANIN
© Armand Colin, 2016, Paris
ARMAND COLIN is a trademark of DUNOD Editeur – 11, rue Paul Bert – 92240 MALAKOFF

First edition published 2020
by CRC Press
6000 Broken Sound Parkway NW, Suite 300, Boca Raton, FL 33487-2742

and by CRC Press
2 Park Square, Milton Park, Abingdon, Oxon, OX14 4RN

First issued in paperback 2022

ISBN 13: 978-1-03-247447-2 (pbk)
ISBN 13: 978-0-367-26129-0 (hbk)
ISBN 13: 978-0-429-29196-8 (ebk)

DOI: 10.1201/9780429291968

Typeset in Palatino
by codeMantra

# Contents

# PART 2    Cartographical Language

# PART 3   Beyond the Visual Variables

# Foreword

In the geospatial domains, we can witness that more spatial data than ever is produced currently. We need to make more and more efforts to deal with all those data in an efficient sense, mining the relevant information and link and select the appropriate information for a particular scenario.

However, how can we unleash the big potential of spatial data in truly interdisciplinary approaches better? How can we make sure that spatial data is applicable for governments, for decision-makers, for planners, and for citizens in an easy-to-use and efficient manner, so that the human user benefits?

In this respect, maps and cartography play a key role. Maps are most efficient in enabling human users to understand complex situations. Maps can be understood as tools to order information by their spatial context. Maps can be seen as the perfect interface between a human user and all those data, and thus they enable human users to answer location-related questions, to support spatial behavior, to enable spatial problem solving, or simply to be able to become aware of space.

Today maps can be created and used by any individual stocked with just modest computing skills from virtually any location on Earth and for almost any purpose. In this new mapmaking paradigm, users are often present at the location of interest and produce maps that address needs that arise instantaneously. Cartographic data may be digitally and wirelessly delivered in finalized form to the device in the hands of the user or he may derive the requested visualization from downloaded data in situ. Rapid advances in technologies have enabled this revolution in mapmaking by the millions. One such prominent advance includes the possibility to derive maps very quickly immediately after the data has been acquired by accessing and disseminating maps through the Internet. Real-time data handling and visualization are other significant developments as well as location-based services, mobile cartography, and augmented reality.

While the above advances have enabled significant progress on the design and implementation of new ways of map production over the past decade, many cartographic principles remain unchanged, the most important one being that maps are an abstraction of reality. Visualization of selected information means that some features present in reality are depicted more prominently than others, while many features might not even be depicted at all. Abstracting reality makes a map powerful, as it helps humans to understand and interpret very complex situations very efficiently.

However, as geodata and mapmaking software instruments become available to many, the need for understanding background, fundamentals, and methods of successful cartographic modeling processes raises significantly. It is rather interesting to witness how many maps are produced nowadays lacking fundamental rules of cartographic practice, thus demonstrating a lack of knowledge.

It is this context which makes every contemporary handbook or manual on cartography highly relevant. What is the fundamental knowledge I should have when dealing with maps, when trying to depict spatial information by graphical means and convey a story, a message for the benefit of human map users? Can there be a manual that describes all the theoretical and methodological underpinnings of cartography but gives at the same time easy-to-understand examples as well as discussing on a holistic level the consequences, limitations and constraints of maps and cartographic processes?

Of course, there are several textbooks, guidebooks, and introduction materials to cartography available, some of them quite old and some of them more technologically focused. The excellent *Manuel de cartographie* published in 2016 by Nicolas Lambert and Christine Zanin in French offered everything needed to gain a mutual understanding of modern cartography. It is a real benefit to the non-French-speaking world that this book is now available in English as well.

It demonstrates a holistic understanding of cartography in the tradition of the famous "Sémiologie Graphique" from Jacques Bertin. This book is divided into a logical "stack", starting with fundamentals on the "input" for mapmaking (basemaps and statistical data), on the methods and concepts of transforming data into graphics and finally, in a more critical approach on looking beyond the visual variables, thus giving insight on how to go even further in cartographic design. With this, interested readers get something like a profound compass in their hand, which can guide them through the amazing world of cartography and help to ensure, that not only spatial data and geotechnologies become available on the fingertips of many but also the theoretical and methodological underpinnings of cartography. As a result, I expect many better maps being produced in future!

As former president of the International Cartographic Association, I claimed always that

1. Cartography is relevant
2. Cartography is modern
3. Cartography is attractive

and that it is therefore "OK, to be a cartographer!".

This book demonstrates why cartography is relevant, modern, and attractive. *Félicitations*, Christine et Nicolas!

Georg Gartner, Vienna

# Authors

**Nicolas Lambert** is a research engineer in geographic information sciences at the CNRS (French National Centre for Scientific Research). Passionate about cartography and dataviz, he has made this activity his core work. He designs geographical maps to decrypt the world but also "protests" maps to try to transform it. As a political and associative activist, he has been involved for nearly 10 years in the cause of migrants within the Migreurop network. He regularly shares his maps and works on the blog neocarto.hypotheses.org. He is known on twitter as "cartographe encarté" (@nico_lambert).

**Christine Zanin** is a professor and researcher in geography and cartography at the University of Paris Diderot and the UMR Géographie-Cités. Her passion for maps and the graphic world leads her to think of a pedagogy committed to cartographic design that respects visual rules but is in search of the innovation made possible by new digital tools. How to think about the spatial organization of territories through the cartographic prism is the meaning of her research.

Nicolas Lambert and Christine Zanin have been working together for more than 10 years to advance cartographic expression in all aspects of their professional or personal commitments. They were jointly awarded the 2009 Paris Diderot University Innovation Prize and published in 2016 the *Manuel de Cartographie* in French to understand and apply the different ingredients for effective mapping. In 2019, they also published *Mad Maps*, an atlas of 60 unpublished maps for the general public to help disseminate mapping.

*A map of the world that does not include Utopia is not worth even glancing at, for it leaves out the one country at which Humanity is always landing. And when Humanity lands there, it looks out, and, seeing a better country, sets sail. Progress is the realisation of Utopias.*

Oscar Wilde, *The Soul of Man under Socialism*

# General Introduction

"La trahison des images" (The Treachery of Images), one of René Magritte's most famous works, painted in 1929, depicts a realistic-looking pipe in profile, resting on a caption painted in the picture "Ceci n'est pas une pipe" (This is not a pipe). With this picture, Magritte wanted to demonstrate that, even if it is the most faithful representation of a pipe, its representation is never really a pipe: it cannot be smoked; it cannot be touched or handled. According to Magritte, a picture is always a representation of an object, but not the object itself. This painting is therefore not reality, but a representation of it, according to hypotheses, intentions, and technical know-how.

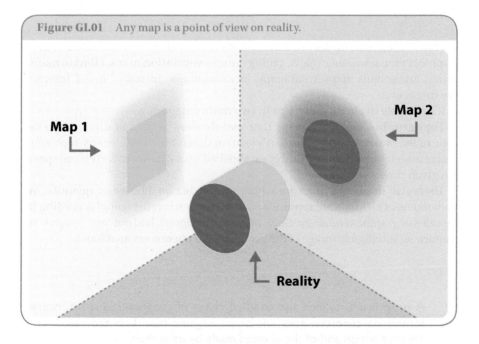

**Figure GI.01**   Any map is a point of view on reality.

Long considered as an impartial, neutral scientific representation of geographical space, "a map is not a territory" (Korzybski, 1931). Resulting from choices and simplification, a map is "a" representation of reality and not reality itself. It is "the production of something concrete and unfamiliar that is converted into an abstract representation" (Lacoste, 1976). Indeed, a map is the "result of a set of technical gestures materializing an idea" (Jacob, 1992). All maps are therefore subjective. They offer a particular proposition, and they do not reflect the world objectively (Brotton, 2014). The possible designs are infinite.

● **FOCUS: The Cartographer's Eye**

Behind a map, there is always a cartographer. If you have 100 cartographers on the top of Empire State Building and ask them to map Manhattan, you will get 100 different map representations. The discrepancy between reality and the image representing it is inevitable. It is all the more true when the phenomena to be mapped are not visible to the naked eye, as it is often the case with human geography. A cartographer is therefore someone who, through his eye, shows to our eye what cannot be seen, a magical eye capable of making the invisible, visible.

## I.1 WHAT IS A MAP?

The term "map" can be declined in several entities. There is a multitude of geographical maps: weather maps, geology maps, vegetation maps, climate maps, tourist attractions maps, road maps, location maps, animated maps, interactive maps, etc.

It's possible to classify all maps in two main categories:

**Topographic maps** are maps that mainly show results of direct observation: relief, water courses, human constructions, etc. These maps represent concrete elements that are durably established on a portion of terrestrial space at a given time.

**Thematic maps** are maps on which localizable qualitative or quantitative phenomena of all kinds are represented. Information is represented according to the rules of graphic semiology. These maps are designed, laid out, and "staged" to produce an intelligible image of the geographical phenomenon at hand.

### Definitions

A **map** is a simplified and codified image of geographical space, representing its characteristics and/or its organization. It is the result of a creative action and of the choices made by its author.

**Cartography** is an artistic, scientific, and technical discipline that aims to conceive and draw up maps.

A **cartographer** is someone who masters the methods, techniques, and concepts of converting geographical information into images.

## 1.2  WHY DO WE HAVE MAPS?

Maps are the starting point and culmination of a geographer's work. They are his specific tools. To put it simply, maps are used to understand space, physical, and human organizations, etc. They enable information to be spatially represented, visualized, and analyzed. Maps allow discovering territories, carrying out and testing hypotheses, and learning about spatial organization. But above all, they are used for communication. Conceived as a way to communicate, they let to transmission of information or ideas, and help to understand geographical analysis.

Geography textbooks are full of maps giving students explanations about different phenomena. The more pedagogical is a map, the more effective it is. All in all, a map serves to tell us about a territory. For it, and to do this, we need to choose words, colors, and curves for the layout, and to adapt to the population targeted. Each map should be designed specifically for a particular group of people and should expound a subject that is thoroughly thought-through.

● **FOCUS: Where Does the Word Cartography Come From?**

The word "cartography" stems from the Greek word *khartes* and the Latin word *carta*. It comes from the designation of the medium that was used: parchment. More recently in Europe, it was the German geographer, Karl Ritter of the Berlin Geographical Society, who used the word *Kartograph* for the first time, in 1828. A year later, the French Cartographic Society followed suit by using the word *cartographie*. In 1859, the British also took over the word "cartography". In addition, the word "map" comes from *mappa*, which is a piece of fabric. It should also be noted that the word "chart", phonetically close to the French word *carte*, refers more to a statistical nonspatial representation (pie chart, diagram).

## 1.3  WHAT IS THE PURPOSE OF THIS HANDBOOK?

This handbook is about thematic cartography. It has been designed as a practical and fully illustrated tool for students to use in geography or infographics. It contains more than 130 figures. It can be read from beginning to end like an essay or read by dipping into it for the information needed.

● **FOCUS: Cartography, Mapping, or Geoviz?**

These three words seem to mean the same thing and seem to be used to describe the process of designing maps nowadays. However, their meanings are not exactly the same. The term "mapping" refers more to the world of GIS and to the display and overlaying of layers of geographic information. The word "cartography" is most often used to designate a meaningful representation (map with message) by combining geographical and statistical data (e.g., thematic or statistical cartography). Finally, the term "Geoviz" (for Geovisualization) always requires a digital and interactive support. Three words, three different worlds.

This handbook is composed of three parts. Part 1 gives details on how to use, construct, and manage the raw material at the cartographer's disposal, i.e., basemaps (Chapter 1) and statistical information (Chapter 2).

Part 2 looks more specifically at the methods and concepts of transforming data into graphic form. The representation of different types of data is introduced: nominal qualitative data (Chapter 3), ordinate data (Chapter 4), and absolute quantitative data (Chapter 5). In addition, Chapter 6 gives information on the use of cartography in particular contexts: temporal data, comparisons, typologies, multivariate analyses, etc.

And finally, in a more critical approach, Part 3, entitled "Beyond visual variables", gives details on how to go even further in cartographic design. Innovative cartography methods are presented (Chapter 7). The choice of the layout and general "staging" (in a theatrical sense) of the map are described in detail (Chapter 8). To conclude, Chapter 9 demonstrates the subjectivity of cartographic images enabling the description of geographical space.

With this handbook, we hope to rehabilitate the term "cartography". Often taken, wrongly, as old-fashioned, in favor of the more contemporary term "geomatics", cartography has nonetheless some advantages. It is a rich and abundant discipline. When we design maps, we become at the same time explorers and pedagogues. Spatial structures are explored, we strive to understand how geographical space works, we analyze it, and we dissect it. Once the analysis has been carried out, results have to be shared. To do this, cartographers design images to tell us about territories. They formalize and depict space and build a world by materializing the imperceptible.

Any map is an invitation ... to see, dream, think, and act. So let us dream, imagine, create, and make maps. And proclaim its legitimacy!

## MAP GAMES

At the end of every section, there are exercises to put into practice the notions that have been presented. With what you learn in the different stages described in this handbook, the objective is step by step to enable you to design a map in accordance with your own particular choices. Your homework will be to produce an interesting and communicative image. The imposed theme, in this game, is the distribution of the US population, but it can naturally be applied to any spatial theme.

The game consists of three parts.

**Part 1:** You will focus on acquiring a coherent set of data, structured and exploitable in a mapmaking approach (basemap, statistical data). In this part, it is important, if required, to process and convert the acquired data (generalization of basemaps, data discretization, etc.)

**Part 2:** You will produce the chosen map using graphic ways suited to the nature of the data acquired during Part 1.

**Part 3:** This concerns the final development phase of the map: you will choose the layout and the "staging" of the cartographic image.

Be aware that the proposed theme (or any other theme) can be addressed in a number of different approaches you are free to choose from total population, density, young people, the elderly, men, women, demographic projections, life expectancy, birth rate, death rate, etc. You also have a free choice of techniques and software you are used to.

**A BRIEF HISTORY OF CARTOGRAPHY...**

> *"Writing history makes a mess of geography!"*
> Daniel Pennac, *La fée carabine*, 1987

...just to have some important cartographic steps in mind!

## THE EARLY MAPS

Geography precedes history. To appropriate their environment, humans have long sought to represent the space around them. Certain maps actually pre-date the invention of writing.

| | |
|---|---|
| 2200 BCE | **The Nuzi tablet**<br>On one of the Nuzi tables found in Mesopotamia near Kirkuk is a map engraved in clay. It is the oldest map ever found. |
| 2000 BCE | **The Belinda map**<br>A map found on the walls of Belinda cave in northern Italy. |
| 600 BCE | **The Babylonian map of the world**<br>A map engraved on a clay tablet found near the town of Sippar, southern Iraq. It is the first attempt to map the world. |

## GREEK SCIENTIFIC CARTOGRAPHY

The Greeks were pioneers of scientific cartography, representing the shape of the Earth and inventing systems of projection.

| | |
|---|---|
| 650 BCE | **The Earth is a disc**<br>Thales of Miletus saw the earth as a flat disc floating on water. |
| 550 BCE | **The Earth is a sphere**<br>For Pythagoras, the Earth was necessarily the most perfect geometrical shape – a sphere. |
| 500 BCE | **The first map of the Ecumene**<br>Anaximander and Hecataeus drew up the first map of the known world, seen as a flat disc centered on the Mediterranean. Three continents can be seen: Europe, Asia, and Africa (Libya). |
| 200 BCE | **The first estimation of the circumference of the Earth**<br>Eratosthenes, curator of the Great Library of Alexandria, performed the first measurement of a meridian arc and made the first estimate of the circumference of the Earth. |

| 120 CE | **Ptolemy's geography** |
|--------|-------------------------|
|        | Ptolemy is considered to be the last great academics of Antiquity. In his *Geography*, he compiled all the geographical knowledge of the time. This work includes 27 maps: 12 maps of Asia, 10 of Europe, 4 of Africa, and 1 of the world. |

## THE UTILITARIAN MAPS OF THE ROMANS

Drawing away from the Greek scientific approach, the Roman maps were restricted to practical and utilitarian aspects – military maps, property maps, and routes.

| 300 CE | **Peutinger's Tabula** |
|--------|------------------------|
|        | This itinerarium comprises eleven rolled parchments accounting for 200,000 km of road, surviving in a 13th-century copy. It is the ancestor of our road maps. |

## THE MIDDLE AGES

In the Middle Ages, European cartography went through a period of virtual vacuum. The medieval mapmaker seems to have been dominated by the church, reflecting in his work the ecclesiastical dogmas and interpretations of scripture. Maps were more symbolic than geographical. The church adapted maps to a mystical and contemplative ideology.

| From the 8th century | **T-O maps** |
|----------------------|--------------|
|                      | The cartographic representations of the Earth to the form of a T inside an O. Oriented east, these maps return to the three continents (Europe, Asia, and Libya) discovered by the Greeks. They are separated by two perpendicular rivers: the Nile and the Don (Tanais). |
| 1240 | **Ebstorf's world map** |
|      | The Ebstorf mappa mundi divides the world into three zones corresponding to the sharing out of the world by Noah's sons after the flood. In the center of the map, we find Jerusalem. Christ dominates the top of the map. On this particular map, on account of the complexity of the information included; the T-O structure is less prominent, although still present. |

## THE ARAB AND ASIAN WORLDS

While scientific mapping was set aside in the West, it re-emerged from the 8th century in the Arab world. By re-appropriating Greek geography, the Arab academics renewed the somewhat forgotten discipline. Arab maps served as a link between West and East.

| | | |
|---|---|---|
| 950 | **The Istakhri map** | The Istakhri map, centered on the Moslem world, is oriented southwards towards Mecca. Here, Europe is reduced to a triangle to the bottom right. |
| 1154 | **The Al-Idrisi world map** | Al-Idrisi was the geographer and physician of king Roger II of Sicily, and was also a great traveler. He visited China, Tibet, and Europe as far as Scandinavia. Although the maps are circular, for Al-Idrisi the Earth was indeed spherical. His maps served as a basis for the first marine maps (portolan charts). |
| 1402 | **The Kangnido map** | Korean world map where European and African continents are designed. China and Korea are oversized. Unlike European or Arabian maps, the Kangnido map presents a square shape. |

## THE REVIVAL OF CARTOGRAPHY
## (THE GREAT DISCOVERIES)

From the 13th century, mapmaking saw an unprecedented rise, at a time when the conquest of new territories primarily required a mastery of sea navigation. The maps were now oriented towards the magnetic north indicated by the compass. Progress in technologies such as navigation, ship design and construction, instruments for observation and astronomy, and general use of the compass tended continuously to improve existing map information, as well as to encourage further exploration and discovery.

| | | |
|---|---|---|
| 1290 | **The Carta Pisana** | The Carta Pisana is the earliest marine map known. It is the first of a series of maps produced from the 13th to the 18th century – the portolan charts. On these maps, only coastal cities are marked. Portolan charts are based on the compass rose or wind rose, and on rhomb lines, and they have no reference system of coordinates. |
| 1492 | **The first terrestrial globe** | The first known terrestrial globe was made by Martin Behaim in Nürnberg. |

| 1507 | **The first map of America**<br>The Waldseemüller planisphere is the first map to show the word "America". |
| 1524 | The first edition of the Petrus Apianus textbooks of geography illustrated with maps and figures reflecting the general eagerness of the times for learning, especially geography. |
| 1569 | **Mercator**<br>The Mercator map was above all a marine map. It was intended to facilitate the task of navigators in keeping on course. |
| 1570 | **Ortelius**<br>First modern atlas. |

## TOWARDS A GREATER ACCURACY

During the 18th century, cartography was characterized by scientific trends and more accurate details. Factual content was privileged with all the increasing information available, often with explanatory notes, and attempts to show the respective reliabilities of some portions. The new cartography was also based on better instruments, the telescope playing an important part in raising the quality of astronomical observations.

Topographic activity is now reinforced to some extent by increasing civil needs for basic data. Many countries of Europe began to undertake the systematic topographic mapping of their territories. Such surveys required facilities and capabilities far beyond the means of private cartographers who had theretofore provided for most map needs. Originally exclusively military, national survey organizations gradually became civilian in character. The Ordnance Survey of Britain, the Institut Géographique National of France, and the Landes topographie of Switzerland are examples.

| 1761 | **The measurement of longitude.** The clockmaker John Harrison devised the first chronometer to cope with the ocean swell, thus enabling accurate measurements of longitude at sea. |
| 1682 | **A Map of France corrected on orders from the king.** Using recent technical and scientific progress, Picard and La Hire drew up a more precise map of the contours of France. On this map, the kingdom appears smaller than before, to the great despair of Louis XIV. |
| 1760–1815 | **The Cassini Map.** This is the earliest triangulated map of the kingdom of France. It was drawn up on the initiative of Louis XV and occupied four generations of a family of cartographers, the Cassini, who devoted their lives to the establishment of this map, which was of unprecedented accuracy. |

| 1799 | **The meter.** Between 1792 and 1798, Delambre and Méchain measured the distance between Dunkirk and Barcelona along the Paris meridian. This measure was to serve as a universal reference to finally define the standard meter. |
| 1891 | The International Geographical Congress in 1891 proposed that the participating countries collaborate in the production of a 1:1,000,000-scale **map of the world**. By the mid-1980s, the project was nearing completion. |

## PIONEERS OF DATA VISUALIZATION

In addition of the cartographic history, we find some non-geographic representation.

| 1370 | **Nicolas Oresme** represents in graphical form the relationship between two variables and prefigures the first bar charts. |
| 1786 | **William Playfair** invents three types of graphic design: bar chart, curve, and pie charts. |
| 1858 | **Florence Nightingale** shows with a pie chart the cause of the death soldiers during the Crimean War. |

## THE BIRTH OF MODERN THEMATIC MAPPING

The start of the 19th century saw the emergence of autonomous thematic maps, illustrating statistical information of demographic, sociological, or economic nature. Mapmaking was becoming a fully fledged discipline. Thematic mapping is the result of the convergence of classical mapping and data visualization.

| 1826 | **The first statistical map** Charles Dupin produced the first thematic map in history – a figurative map of popular education. |
| 1828 | **The appearance of the term "cartography"** The German geographer Karl Ritter used the word "Kartograph" for the first time. |
| 1869 | **Charles Joseph Minard** represents the colossal losses of the French army in the Russian campaign at the beginning of the 19th century. This famous map is nowadays considered as a "gold standard". |

## CONTEMPORARY CARTOGRAPHY

Today maps are no longer drawn by hand; they are designed on a computer. Maps have diversified – they can be interactive, animated, participative, three-dimensional, etc. As they are easier to produce, they have become part of our everyday lives.

| | |
|---|---|
| 1940 | **The first computer**<br>The mathematician Alan Turing established the theoretical basis for what would become a computer. |
| 1960 | **The quantitative revolution in geography and the first Geographical Information System (GIS)**<br>Drawing on the resources offered by mathematics and computing, geography sought to define itself as a fully fledged science. The year 1960 also saw the establishment in Canada of the first GIS. |
| 1966 | **The first photograph of the earth**<br>This first image of the earth from space was taken from the moon orbit by the Luna Observer spacecraft. |
| 1967 | **The Graphic Semiology**<br>Jacques Bertin published the founding treatise on contemporary cartographic language, *La Sémiologie Graphique*. |
| 2004 | **The launching of Google Maps**<br>While maps were for a long time the prerogative of nation states, today it is an American multinational that offers accurate maps of the whole world. |

# Part 1

# Geographical Information

*After having told young children that the Earth is round, that it is a ball roll-ing through space like the sun or the moon, can I then show them an image in the shape of a rectangular sheet of paper with colored images? [...]Should I try to make these little ones understand that the sphere has been changed into a planisphere (flat world map) – in other words, if I correctly understand the two juxtaposed words, a flat sphere?*

Elisée Reclus, *Learning Geography,* 1903

## INTRODUCTION

Data mobilized in geography and cartography is referred to as geographical data. This data relates to space and/or to phenomena localized on the sur-face of the Earth. Any information with an address or any element enabling localization in space (e.g., the identification of a place or a landscape) can become a piece of geographical information. Several definitions are pro-posed. According to Michael Frank Goodchild (1997), geographic information is information about places on the Earth's surface, knowledge about where something is, and knowledge about what is at a given location. In summary, it is information that relates to one or more places in the terrestrial space (Beguin and Pumain, 2014). The geographic space is determined by reference to coordinates, either longitude and latitude for conventional mapping or some others forms for distorted maps.

Geographical information makes it possible to orient in space (where is a given phenomenon to be found?) and to compare localizations (why here rather than elsewhere, why here more than elsewhere?). In geography, this characteristic is essential, because a single, isolated piece of numerical data has no real meaning. It takes on meaning in comparison with other localities.

● **FOCUS: Data or Information?**

The terms "data" and "information" refer to concepts that overlap and can appear as synonymous. Data can be stored (figures, written material, photographs, videos, etc.). Information is the result of the action of interpreting data. For instance, gross domestic product (GDP) data give us information on the wealth of a country. It is possible to store the data enabling information to be produced from a database, but not this information itself, which results from a process of construction and interpretation. It is nevertheless common to speak of geographical information when what is meant is geographical data. This handbook is no exception.

To give data a spatial dimension, there are two possible operations:

- *Geo-referencing* consists in positioning or marking by hand an object on a reference basemap. This operation can be conducted with the computer mouse, identifying the place to be referenced by a click, or by entering the geographic coordinates of the object if these are available directly, using the keyboard. Adjusting an image involves positioning exactly an object (such as an aerial photograph), so as to superimpose it on the basemap. Geo-referencing can also be performed directly in the field using a GPS device.
- *Geocoding* consists in attributing geographic coordinates (longitude, latitude) to an address. To do this, a database is required giving references for each section of road, along with tools to exploit the database. These geocoding systems, which were for a long time on offer on the market, are now available, free of charge. For the USA, the website https://www.census.gov makes it possible to geocode addresses across the national territory via an online application.

It is often said that there are two types of geographical information:

- *Reference information* concerns general data that can be used in numerous areas of activity: administrative boundaries, road networks, countries across the world, etc. A lot of this information is freely available online.
- *Thematic information* concerns particular themes. It can be produced by businesses, study bureaus, or specialists of one or other domain. This information is more specialized and generally come at a cost. Today, the open data movement is facilitating access.

● **FOCUS: Open Data**

Open data is a movement that started up in the USA in the 1990s. It expanded in the 2000s with the development of Internet. It consists in issuing data (geographical or other) with an open license guaranteeing free access and use of the data. Today, many countries and cities around the world have switched to Open Data, which greatly facilitates the work of cartographers.

Whether the geographical information you are to use concerns reference information or thematic information, it can be divided into three types:

- **Geometric information**, formed by points, lines, and polygons, describes the shape, the contour, and the localization of objects within a system of geographical or Cartesian coordinates. This is the basemap (Chapter 1).
- **Semantic information**, made up of numerical values or text, relates to an object that is localized in space. It can be the name of a street, a plot number, or statistical information concerning the particular territorial unit (e.g., the number of inhabitants in Cherokee County (North Carolina). This type of data is often known as attribute data (Chapter 2).
- **Topological information** is information deduced from the geometrical information. It is defined by spatial relationships between objects: proximity, distance, contiguousness, inclusion, etc. It can generate specific cartographic representations such as discontinuities, multi-scalar typologies, or smoothing.

Any thematic map will be the result of a combination of a basemap (geometrical information) and one or several types of data (semantic information).

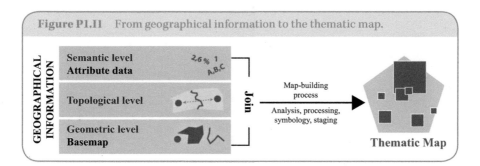

**Figure P1.I1**    From geographical information to the thematic map.

Geographical information is the cartographer's raw material. Making a map means converting geographical information into an image. Which basemap should be used, what cover, what level of generalization, which type of projection, which grid? What is the best statistical indicator to use, how should it be expressed, should you use raw or converted data? These initial choices are all part of the process of designing a map. The first part of this handbook will seek to describe in detail the two main components of geographical information, essential for the design of thematic maps: geometric information (the basemap) and semantic information (attribute data).

# Chapter 1

# Geometrical Data

▸ **Objectives**

- Choosing and/or creating a basemap
- Knowing the principles of generalization
- Knowing the differences between a vector image and a raster image
- Understanding the issues raised by cartographic projections
- Understanding the impact of spatial subdivisions on cartographic representations.

Choosing a basemap, whether to be developed from scratch or merely downloaded from the Web, is an essential step in cartographic creation. The basemap is what will contain the information that is to be represented. It is the base that enables the spatialization of a phenomenon. It needs to be suited to the complexity of the map to be developed, and to the objectives of the cartographic project.

## 1.1 GEOGRAPHIC OBJECTS

A basemap is made up of a set of geographic objects that need to be chosen carefully, according to the level of detail required, the medium used for the representation (paper or digital), its purpose, and the type of public targeted. There are three types of geographic objects (three basic graphics): the point, the line, and the polygon. These geographic objects can be constructed in two modes: vector and raster (matrix). They are linked to one another via a spatial relationship (proximity, distance, contiguity), by their size and by their shape. These objects can be geo-referenced and stored in coherent manner in a geographic information system (GIS). They form layers that can be superimposed, crossed, and used to develop a basemap. They enable the reality of the world to be "drawn" or depicted on a flat surface (a sheet of paper or a screen).

In thematic mapping, vector basemaps are the more frequently used. Unlike the raster format, the vector format enables a single identifier to be associated with each object, which in turn enables these objects to be linked individually to statistic tables using the same codes. This operation, which is widely used in cartography, is known as the **join**. Once the map is completed, it can easily be revised using CAD (computer-aided design) software. It is thus possible to alter, re-size, or complete the image without pixelization or loss of graphic quality (see Chapter 8).

## Definition

**Join:** a join consists in the linking of statistical information to an object that is already localized, by confronting two identical codes.

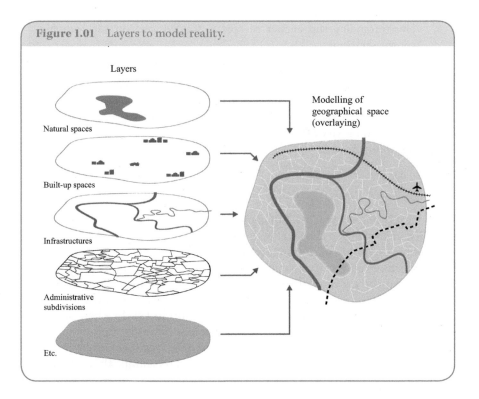

**Figure 1.01**     Layers to model reality.

Layers

Natural spaces

Built-up spaces

Infrastructures

Administrative
subdivisions

Etc.

Modelling of
geographical space
(overlaying)

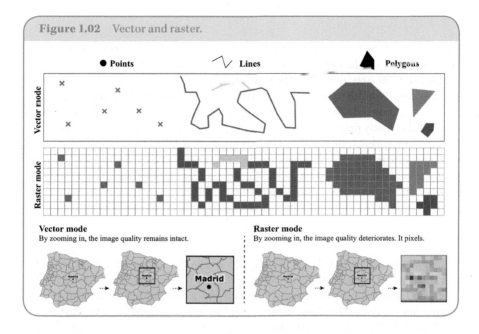

**Figure 1.02**   Vector and raster.

A basemap is determined by four elements: the way in which the space represented is laid out (the **projection**), the choice of the portion of terrestrial space considered (the **spatial extent** and the **orientation**), the way in which the basemap is drawn (the **generalization**), and the geographic objects represented on it (the **spatial subdivisions**). The choice of these elements is essential in map design. Although in practice the cartographer will not create a new basemap for each new map, he or she will nevertheless need to be realistic and critical with regard to the materials to be used: why this basemap rather than another? Can it be improved? What impact does it have on the message to be conveyed by the map?

Indeed, while the basemap is a mere "container", it is also the framework and the medium of the geographic information to be represented. Information is laid out on it, and at the same time, it is itself a part of the information. Drawing dividing lines and borders has an impact on the visual quality of the transcription of the message and hence on its efficacy. Choosing specific spatial subdivisions on which the information is to be described, and deciding whether it is suited not only to the data but also to its representation are the first issues to be considered by cartographers, whether novice or experienced. Below are a few elements to help in these decisions.

## 1.2 PROJECTIONS

Whatever the base medium used (a sheet of paper, a computer screen, or a tablet), the world has to be laid out flat, or "projected" in order to be seen in its entirety.

Although for Aristotle, the area covered by "dry land" and inhabited by man was sufficiently small to be assimilated to a plan (Claval, 2011), the mathematical process that consists in converting a three-dimensional space into something that is flat already preoccupied the Greek geographers of Antiquity. This operation involves several constraints: there is no ideal projection, and any projection produces distortions. Two basemaps that do not have the same projection cannot be correctly superimposed. The larger the space represented, the greater the distortions. This conversion operation enabling the shift from the spherical earth to the Cartesian plane (the basemap) is conducted in several stages.

● **FOCUS: How Long Have We Known That the Earth Is Not Flat?**

Despite what is often said, humanity has known for a long time that the Earth is not flat. In *The Geography*, Ptolemy (90–168 CE) already gave a description of two projections that could be used for a map of the ecumene (areas that are known, inhabited, or exploited by man). Before him, Eratosthenes (276–194 BCE) calculated one of the first estimations of the terrestrial circumference, 39,375 km, which is very close to reality (40,075 km).

### 1.2.1 Finding a Localization on the Surface of the Earth

The Earth can be assimilated to a sphere, but it is not a perfect sphere. There are mountains and ocean trenches, so that the Earth is irregular in shape. To obtain a simplified representation of the terrestrial surface, we need a theoretical intermediate surface, which is known as the geoid and corresponds to a measure of the Earth's field of gravity. The **geoid**, which coincides with the average ocean surface, provides a precise representation of the Earth's surface without the relief. It is the surface area at altitude zero. But far from being a perfect sphere, the geoid is itself distorted. It is flattened at the poles, and different points on the surface of the geoid are not all at the same distance from its center.

Starting from the geoid, it is possible to define the most representative **ellipsoid**. This is a simple mathematical form that is the closest possible to the geoid. The ellipsoid can be global (fitted to the terrestrial globe) or local (so as to fit the shape of the Earth at a given location as accurately as possible). Once assimilated to an ellipse, the surface of the globe can be subdivided by a network of lines enabling each point in space to be localized. From south to north, parallel lines indicate latitude, and from east to west, the meridians define longitude. Any place can thus be localized by way of its geographic coordinates in terms of latitude and longitude.

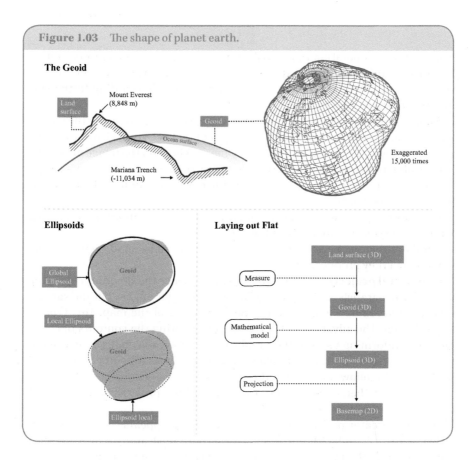

**Figure 1.03**   The shape of planet earth.

**The Geoid**

Mount Everest
(8,848 m)

Land
surface

Geoid

Ocean surface

Mariana Trench
(-11,034 m)

Exaggerated
15,000 times

**Ellipsoids**

Global
Ellipsoid

Geoid

Local Ellipsoid

Geoid

Ellipsoid local

**Laying out Flat**

Land surface (3D)

Measure

Geoid (3D)

Mathematical
model

Ellipsoid (3D)

Projection

Basemap (2D)

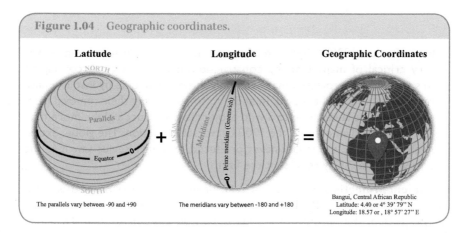

**Figure 1.04**   Geographic coordinates.

**Latitude**

NORTH

Parallels

Equator    0

SOUTH

The parallels vary between -90 and +90

**+**

**Longitude**

Meridians    Prime meridian (Greenwich)

0

The meridians vary between -180 and +180

**=**

**Geographic Coordinates**

Bangui, Central African Republic
Latitude: 4.40 or 4° 39' 79'' N
Longitude: 18.57 or , 18° 57' 27'' E

## 1.2.2 From Ellipsoid to Flat Surface

Projection is an operation that consists in converting a given ellipsoid into a flat surface. This mathematical operation can be performed in a number of ways. Thus, cartographic projections can be classified according to the way they are constructed.

**Azimuthal projections** are constructed by projecting points on the sphere onto a flat surface. The position of the point of contact between the plane and the sphere (center of the projection) naturally has an effect on the image projected.

**Cylindrical projections** are constructed by the projection of points on the sphere onto a tangent or secant cylinder. The developed (or unrolled) cylinder provides the flat surface.

**Conical projections** are constructed in the same manner by projecting points on the sphere onto a tangent or secant cone.

It follows that map projections can also be classified according to the type of distortion they produce.

**Conformal projections** preserve angles (and hence the shape of geographic objects). The parallels and meridians intersect at right angles, but the surface areas are altered as we draw away from the center of the map. This type of projection has the advantage of correctly representing the contours of countries, thus informing users of their real shape. These projections, long used in navigation, also have the advantage of enabling a linear itinerary to be traced on the map to follow a route. On these maps, the distortions of surface area increase the further, one gets from the center of the projection, which is not always the center of the map.

On the Mercator projection, for instance, Russia appears twice the size of Africa, while it is in fact half the size. Even more misleading, Latin America appears smaller than Greenland, while it is in fact nine times larger.

● **FOCUS: Elisée Reclus and His Geographic Utopia**

Elisée Reclus (1830–1905), one of the pioneers of modern geography, was very critical of maps that, by construction, distort reality. According to this author, flat, projected maps can only mislead students by instilling false representations of the world. In addition to advocating their removal from classrooms, he launched into a wild project to construct a globe more than 139.4 yards in diameter for the 1900 Universal Exhibition. The globe, excessively expensive, was in fact never produced, although Reclus saw it as the only way to faithfully represent the Earth.

**Equivalent projections** preserve the relationships between surface areas, but the contours are markedly distorted on the edges of the map. On maps using this projection, the outlines of countries can be difficult to identify, which hampers the reading of the map. Nevertheless, the preservation of relationships between surface areas is valuable in thematic maps wherever the size of a country is relevant to the theme represented (for instance, population density).

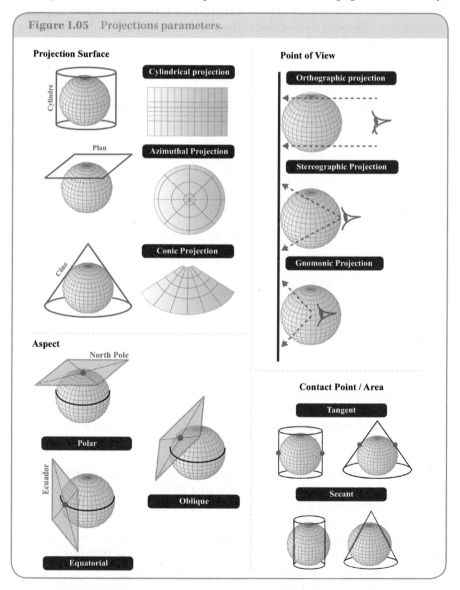

**Figure 1.05**    Projections parameters.

**Equidistant projections** display the actual distances from the center of the projection or along particular lines. But surfaces and shapes are not preserved.

**Aphylactic** – compromise – **projections** do not preserve angles, sizes, or distances, but attempt to compensate for these three distortions by way of a compromise, giving them the name of compromise projections.

The choice of the best projection is a central issue. However, at this stage the cartographer has numerous options. The answer to this is simple: everything depends on the scale!

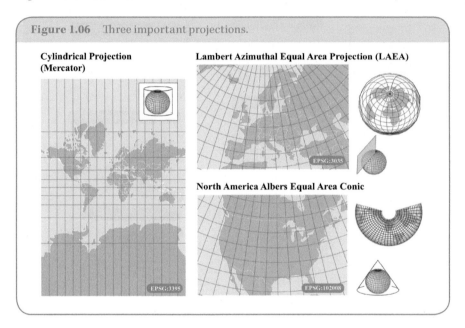

**Figure 1.06**  Three important projections.

For example, In the USA, Albers equal-area conic is the typical projection for historical USGS (United States Geological Survey) map, it being a general-purpose low-distortion compromise for mid-latitude short and wide extents. This conical, conformal projection (retaining the shape of objects) can be implemented in several manners.

In Europe, there is likewise a reference cartographic projection, the Lambert azimuthal equal-area projection. As indicated by its name, it is an azimuthal projection (constructed from a tangent plane) that is also equivalent; that is, it preserves the relationships between surface areas. The drawback of this projection is therefore that it distorts the outlines of countries, but it has the advantage of preserving their surface areas, which is useful in thematic maps. Most of the basemaps issued in Europe use this projection.

To represent the world, there is a wide choice for cartographers. As the space to be represented is vast and cannot be viewed in its entirety because of the curvature of the globe, the cartographer can play on the scope for distorting maps to create one view of the world rather than another. While a ready-to-use basemap with a given projection is most often suitable, a cartographer who is conscious of the need to reflect on the message to be delivered needs to consider the issue of the projection. Is the projection on offer appropriate? Can the map be improved by switching to a different projection? The projection of a map is indeed an efficient means of expressing a message, as Amo Peters showed with his eponymous projection, which raised considerable debate (see "Focus" below).

When choosing a projection and all its attributes like center, edges, or orientation, the cartographer possesses numerous tools to perfect his/her message. The choice of the projection enables certain territories to be enlarged or reduced, which has an impact on the final bounding box of the map, i.e., the spatial extent. The choice of a particular center can be an effective way to express a viewpoint.

**Figure 1.07**    Each projection has its particular message.

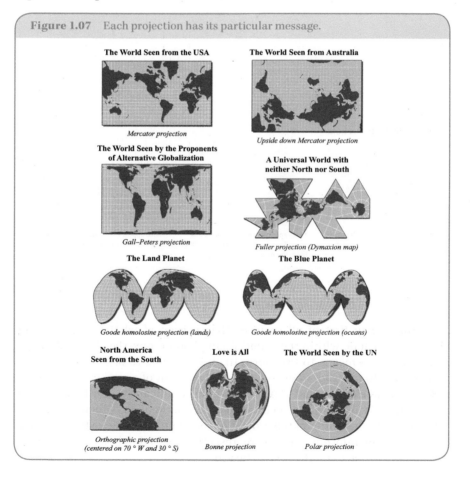

The World Seen from the USA

*Mercator projection*

The World Seen from Australia

*Upside down Mercator projection*

The World Seen by the Proponents of Alternative Globalization

*Gall–Peters projection*

A Universal World with neither North nor South

*Fuller projection (Dymaxion map)*

The Land Planet

*Goode homolosine projection (lands)*

The Blue Planet

*Goode homolosine projection (oceans)*

North America Seen from the South

*Orthographic projection (centered on 70 ° W and 30 ° S)*

Love is All

*Bonne projection*

The World Seen by the UN

*Polar projection*

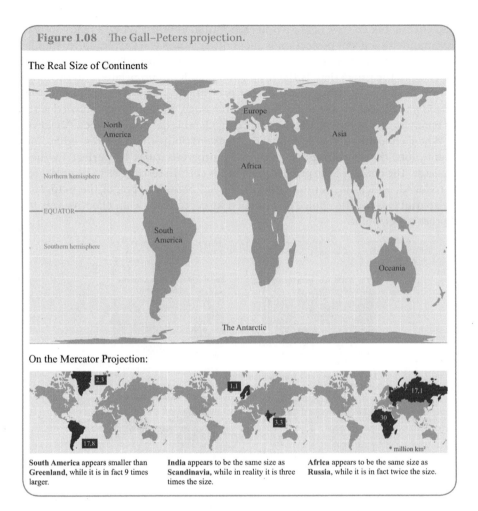

**Figure 1.08**   The Gall–Peters projection.

The Real Size of Continents

On the Mercator Projection:

**South America** appears smaller than Greenland, while it is in fact 9 times larger.

**India** appears to be the same size as **Scandinavia**, while in reality it is three times the size.

**Africa** appears to be the same size as **Russia**, while it is in fact twice the size.

## 1.3  SPATIAL EXTENT AND ORIENTATION

Designing a map first of all requires the decision of what to include and what not to include within the frame. As in photography, the choice of image composition is a means to transform the information, to orient the way it is read, or even to mislead. Choosing the spatial extent of a map can enable the exclusion of an element that, if present, would significantly alter the meaning of the map. It is also a way to define the center of the map (which only rarely corresponds to the center of the projection) or again a technique to maximize the status of the information to be represented. The right choice of cover also enables empty spaces on the map to be avoided.

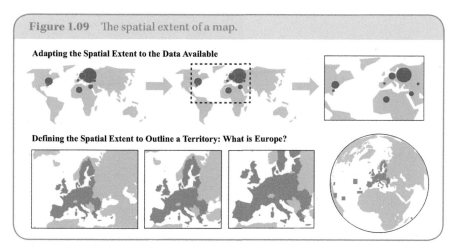

Choosing the spatial extent of a map means choosing where the representation of a territory starts, and where it ends. Should a map of Europe include Turkey, Ukraine, Lapland, or French Guyana? Where is the center of Europe? Should a map of the United States systematically include its remote territories in the Pacific Ocean?

All in all, geographic observation and reasoning are always dependent on the size of the space considered (Lacoste, 1976).

The rotation (or orientation) of the basemap is also a way to influence the area covered and the general aspect of the map. Orientation in relation to the north is not a dogma that cannot be overcome. The habit has indeed evolved over time.

Although in Antiquity, the Greeks were already orienting their maps towards the north, by the Middle Ages, maps were oriented towards the East and the Levant. As for the Arabs, they oriented their maps southwards towards Mecca. Since the age of the great discoveries and the use of the magnetic compass for orientation, the orientation towards the north has taken precedence (Chinese sailors were using the compass by around the year 1000, well before the Arabs and European, and their maps were already oriented in relation to the north).

The orientation of maps in reference to the north is in no way compulsory, this being even more so in the case of thematic maps, where the way in which it is read and the manner in which the information is presented are essential. Nothing ties us to an obligation to comply with an orientation that is determined by the terrestrial magnetic field. While it may be more efficient to use a basemap of the type we are used to dealing with, it can also be interesting to choose an unusual viewpoint to serve an out-of-the-ordinary or controversial message. In this case, the orientation will be indicated on the map by an arrow enabling the reader to position himself. Finally, certain projections like the Fuller projection (see above) have neither north nor south.

**Figure 1.10**  Adapting the geographical extent to maximize the space represented.

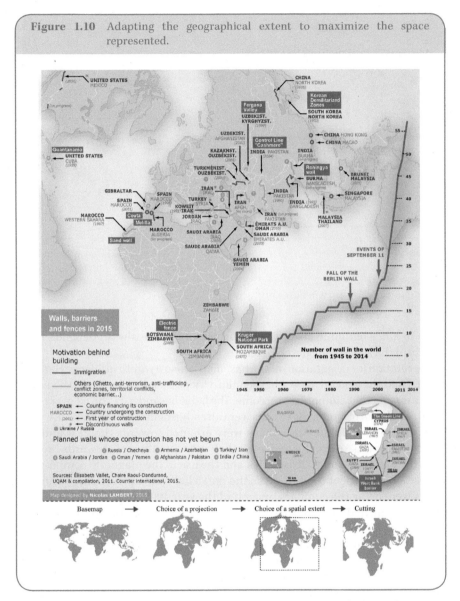

*Ever more numerous walls in a borderless world – this map shows the location of inter-state walls. While there were only a dozen in 1945, there are more than 70 today. The total length of the walls throughout the world reaches an impressive 40,000 km, the equivalent of the earth's circumference. Beyond the projection that enables all the territories concerned by this theme to be seen, the choice of the spatial extent of the map makes it possible to restrict "empty" spaces that are not useful to the understanding of the phenomenon.*

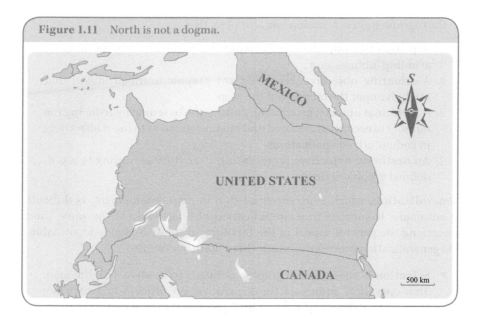

**Figure 1.11**   North is not a dogma.

UNITED STATES

MEXICO

CANADA

S

500 km

● **FOCUS: Did You Know?**

It can be remarked that the word "NEWS" is made up of the letters N (north), E (east), W (west), and S (south).

● **FOCUS: Upside-Down Maps**

In 1943, in order to denounce the Western view favored by world maps, the Uruguayan artist Joaquin Torres Garcia (1874–1949) designed a map of America reversing south and north. According to him, this reversal was needed to shake off the domination of countries in the north. Today, numerous Australian and Chilean world maps are presented "head down", vindicating this artist (e.g., Stuart McArthur's world map in 1979).

## 1.4 CARTOGRAPHIC GENERALIZATION

All maps are reduced reproductions of a portion of space. They therefore require a simplification of the representations. The reduction in scale and the minimum thickness of a line to remain visible on the map force the cartographer to simplify the contour, thus reducing the geometrical precision in order to enhance readability. A map developed with excessive detail in the contours, far from adding precision, risks "obfuscation" when compressed or downsized. This distracts from the main message of the map.

Cartographic generalization has four objectives:

1. A **pragmatic objective:** enhancing the readability of the map by avoiding "obfuscation".
2. A **scientific objective:** removing any graphic overload (noise) that might hamper the message of the map.
3. A **technical objective:** reducing data on the basemap by reducing the number of coordinates stored and memorized so as to use it effectively in computerized applications.
4. An **aesthetic objective:** representing a territory according to a pre-defined graphic style.

**Generalization**, which is an essential step in map development, is difficult to automate. It consists in a simplification of lines while at the same time preserving the general aspect of the territory so that it remains identifiable. The generalization procedure entails three main operations:

- **Selection** is choosing the graphic elements to be shown on the map. This operation addresses the generalization process in two ways: choosing categories of data to be represented (one shows only roads but not railroads) and/or choosing the amount of information within categories (should the outlines of islands or enclaves smaller than a certain surface area be retained?).
- Producing a simplified form: the geometric operation of **simplification** of the lines to be traced (**structural** sketch diagram). This can also be a **conceptual** sketch diagram when it includes elements of interpretation and manual choices (grouping of objects, enhancement, exaggeration, smoothing, grouping, displacement, symbolization, etc.).
- **Harmonization**: this is an operation of general harmonization of the basemap to obtain a comparable level of generalization at any point on the map.

Even if the generalization process can be partly automatic, a well-generalized map is unlikely to be achieved without a manual stage. This is no doubt an advantage, because the basemap and its level of generalization are among the elements available to the cartographer to efficiently express the message intended. It is possible to draw away from the precise localization of objects to facilitate readability of the map. Thus, the cartographer can move certain characteristic lines or forms so as to correct graphic incompatibilities, group certain smaller characteristics into larger units, remove certain objects and enlarge others, or change the graphic representations. Thus, generalization

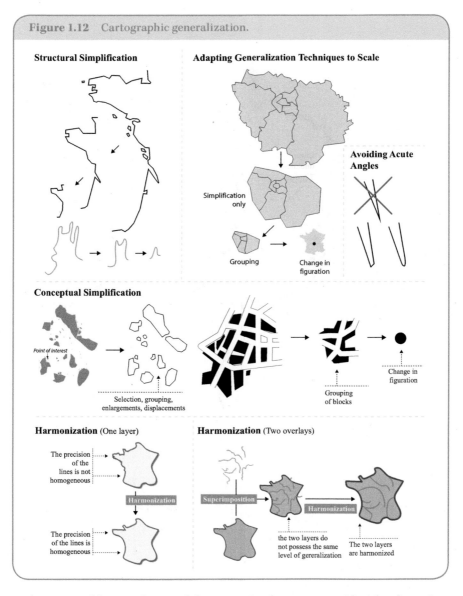

**Figure 1.12**   Cartographic generalization.

makes it possible to reduce or delete certain elements considered to be unimportant, or conversely to focus on certain locations that are important for the intended message. Choosing the level of generalization expresses at least part of the message.

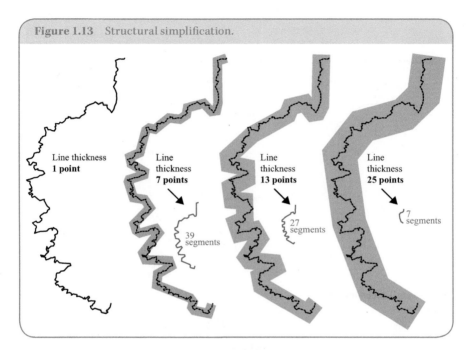

**Figure 1.13    Structural simplification.**

Line thickness **1 point**

Line thickness **7 points**    39 segments

Line thickness **13 points**    27 segments

Line thickness **25 points**    7 segments

*How can a basemap be simplified manually? The colored band shows the generalized line for the western coast of Corsica. By varying the scale and/or the width of the line drawn, the original line (in black) is simplified to a greater or lesser degree. The segments reduce in number with the simplification.*

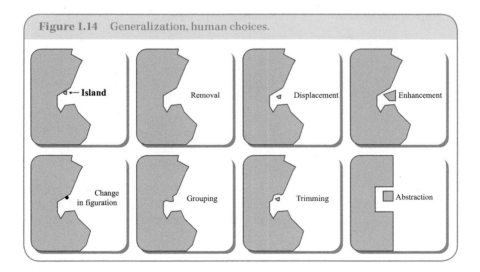

**Figure 1.14    Generalization, human choices.**

← Island

Removal

Displacement

Enhancement

Change in figuration

Grouping

Trimming

Abstraction

**Figure 1.15**  Graphic styles: curves.

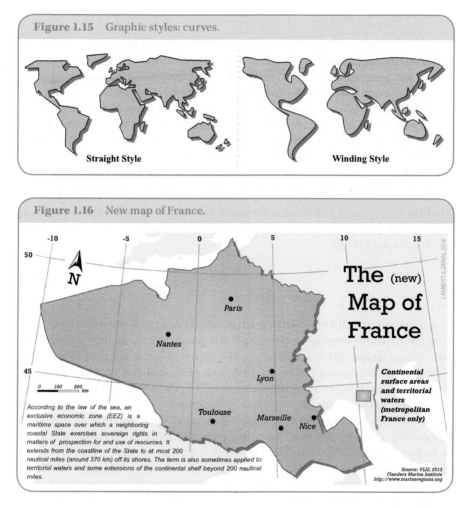

**Straight Style**                    **Winding Style**

**Figure 1.16**  New map of France.

The (new) Map of France

*Continental surface areas and territorial waters (metropolitan France only)*

According to the law of the sea, an exclusive economic zone (EEZ) is a maritime space over which a neighboring coastal State exercises sovereign rights in matters of prospection for and use of resources. It extends from the coastline of the State to at most 200 nautical miles (around 370 km) off its shores. The term is also sometimes applied to territorial waters and some extensions of the continental shelf beyond 200 nautical miles.

Source: VLIZ, 2013
Flanders Marine Institute
http://www.marineregions.org

*Review your geography – France is not what you thought it was! Here is its real aspect....*

*What in fact is lacking on our usual maps? No less than the sea. It is true that few people live on the sea. But does this justify representing only land masses on maps? Maybe but...other spaces on Earth are uninhabited – mountains above the snowline, deserts, and polluted sites – but would we consider removing them from our maps? No, we would not. So why do we wittingly exclude such a large portion of national territory, which for a human population that will soon number nine billion will gain increasing importance? Today we fish in the sea, they are fish farms, we navigate on the sea, we use the sea to produce energy, we transport our goods by sea, we find our raw materials there – and these are all elements of human activity. Why should we amputate France of one of its important components on our maps? So here is the new map of France, as it should be found in geography textbooks. For these are the real boundaries of France, in its terrestrial and maritime dimensions. Something we need to "get our heads around".*

● **FOCUS: The Subliminal Geometry of the Basemap**

Beyond the scientific and technical aspects, the basemap can also conceal a subliminal message. In Patrick Schulmann's French film *P.R.O.F.S.* (1985), the history and geography teacher Charles Max (sic) describes two maps hanged in the classroom: "See how the outline of a State reflects the spirit of the State. A primitive spirit (showing a map of the USA)! Look at these straight lines, arbitrary and reactionary. Here, on the other hand (showing a map of the USSR) look how the borders of the States are fitted to the mountains and the rivers, in other words to man. This shows a spirit of justice and generosity. Even the coastline! See how gentle the borders are (on the map of the USSR). See here (on the map of the USA) how indented and completely aggressive they are".

This film excerpt is a funny illustration of the meaning that can be put in a mere basemap, even if it's done in poor faith. ➲ *https://www.youtube.com/watch?v=58qi6G7lV9E*

## 1.5 THE SPATIAL SUBDIVISIONS

From a geometric point of view, the grid (or spatial subdivisions) refers to a strict partition (without overlap or blurring) of a geographic area into contiguous units whose shape and size can be regular or irregular. It is a partition of the space whose elementary pieces are polygons of shapes and surfaces that are often heterogeneous, interlocking each other like puzzle pieces (John Spilsbury, the inventor of the puzzle in 1766 was a geographer).

Beyond the mere administrative subdivision of space, spatial subdivision systems are also (and above all) reading grids or subdivisions enabling a reality to be captured. They provide a mesh or "canvas" for comprehension and analysis, based on the simplification and generalization of information. They are tools for constructing knowledge. Grids are territorial filters in which each unit provides information for the understanding of the spatial phenomenon analyzed. Thus the statistical, cartographic and modeling results depend on the grid or type of subdivision chosen. A change in size, shape, or positioning of the different elements, even if minute, will affect not only the statistical results, but also their cartographic representation.

Several studies, referred to the acronym MAUP (Modifiable Areal Unit Problem), show that the subdivision of space adopted affects the reading of the spatial organization of a phenomenon, sometimes very markedly. The size of the territorial units also has an impact on the reading of the information, in particular because certain statistical processing operations, upstream of their representation, are directly influenced by this parameter. For instance, the correlation between two phenomena can be significant using one areal unit, but not using a different one. Likewise, if a certain form of spatial organization can be perceived on one map using a given unit, it will not necessarily appear using another. The use of a subdivision made up of administrative units of very different sizes is genuinely problematic: it amounts to constructing a map representing non-comparable geographic units in a single image.

● **FOCUS: The MAUP**

The MAUP concept was proposed by Openshaw (1978) and Taylor (1979) to refer to the influence of spatial subdivisions (scale effects and zone effects) on the results of statistical calculations or modeling procedures.

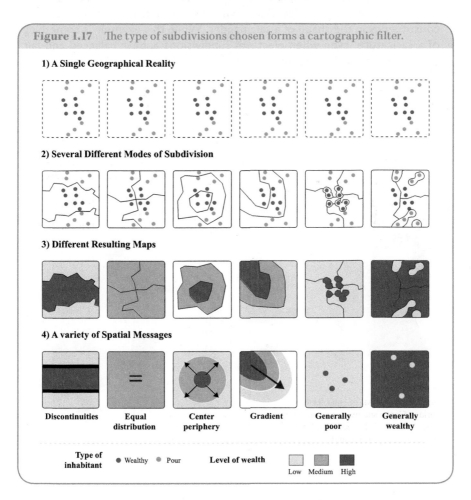

**Figure 1.17**    The type of subdivisions chosen forms a cartographic filter.

1) A Single Geographical Reality

2) Several Different Modes of Subdivision

3) Different Resulting Maps

4) A variety of Spatial Messages

| Discontinuities | Equal distribution | Center periphery | Gradient | Generally poor | Generally wealthy |

Type of inhabitant   ● Wealthy   ● Pour     Level of wealth     Low   Medium   High

It is therefore fundamental to take this into account when interpreting cartographic images derived from these irregular subdivisions (to manage the MAUP issue, cartographic solutions are proposed in Chapter 7 of this handbook).

● **FOCUS: Subdivision and Gerrymandering**

Far more than a means of perceiving and analyzing spatial structures, the system of subdivision used is also part of a power game. The person possessing the power to define the subdivision of space can draw advantage from it. For instance, the North Americans can indulge in gerrymandering. This practice consists in subdividing the electoral constituencies in such a way that a particular candidate is favored. The term is derived from the name of Governor Elbridge Gerry, who in 1811 plotted out an electoral constituency in the shape of a salamander. While this example is something of a caricature, even when electoral subdivisions appear balanced and fair they are never without significance.

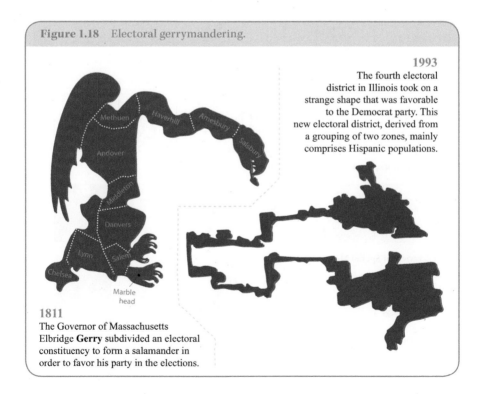

**Figure 1.18**    Electoral gerrymandering.

**1993**
The fourth electoral district in Illinois took on a strange shape that was favorable to the Democrat party. This new electoral district, derived from a grouping of two zones, mainly comprises Hispanic populations.

**1811**
The Governor of Massachusetts Elbridge **Gerry** subdivided an electoral constituency to form a salamander in order to favor his party in the elections.

## Quiz

- What is an ellipsoid?
- What is the MAUP?
- What are the three operations involved in generalization?
- What is an equal-area projection? A conformal projection? An aphylactic projection?
- Why are maps oriented in relation to the north?
- Why are the differences between vector and raster?
- Why does the choice of the spatial extent have an effect on the representation?

# Chapter 2

## Statistical Data

**Objectives**

- How to construct a data table
- How to identify the nature of statistical indicators
- Getting to know the different discretization methods
- Choosing the right discretization method.

Statistical data is either quantitative or qualitative, either collected or constructed, enabling a cartographic representation. We consider that this data is the **semantic dimension** of geographical information. While the basemap can be seen as the "container", the data is the "content". It forms the base of the geographical information represented and delivers the geographical message by way of spatially organizing that what is revealed by the map. This is the data to which the rules of symbolization (graphic semiology) apply (see Part 2).

## 2.1 DATA TABLES

Statistical data is stored in the form of elementary tables in which each line corresponds to a spatial unit (or a geographical unit) and each column corresponds to a variable (or an indicator) that characterizes the object. The data is captured, modified, and handled by way of a spreadsheet. Producing a map requires these data tables to be very thorough and accurate so as to avoid any ambiguity when integrating the data into the basemap.

The first line in the table identifies the names of the different variables. They should be as short and explicit as possible. For reasons of compatibility with certain software programs, it is preferable to avoid special characters, spaces, and accents. The first column serves to identify each single territorial unit by a specific code. These codes are often related to coding systems from the different data suppliers (US Census Bureau, Eurostat, World Bank, etc.). However, in the case of a completely new dataset an intelligible coding system will need to be devised. The units or "boxes" in the table formed in this manner correspond to the different values taken on by the spatial units for each of the variables.

Data tables can include missing values. These can be referenced by way of an empty cell in the table, or by "NA" or "N/A", meaning not attributable or not available. In certain tables, for historical reasons relating to the data formats in some software, missing data are sometimes coded -9999. To avoid any risk of confusion, this practice should be avoided. Thus, missing data will appear as blanks on the map and be referenced in the legend as such.

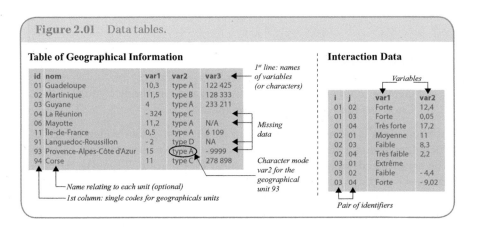

**Figure 2.01**   Data tables.

● **FOCUS: What about Nomenclatures?**

A nomenclature is defined like a set or system of names or terms, as those used in a particular science or art, by an individual or community. We can call *countries nomenclature* the country codes which are short alphabetic or numeric geographical codes (geocodes) developed to represent countries and dependent areas, for use in data processing and communications. Several different systems have been developed to do this. The term "country code" frequently refers to ISO 3166-1. However, each international organization has their own countries codes. For example, WIS codes for the World Bank and M49 Standard "Standard Country or Area Codes for Statistical Use" for the United Nations system.

The ISO 3166-1 system SO 3166-1 is part of the ISO 3166 standard published by the International Organization for Standardization (ISO). The alphabetic country codes were first included in ISO 3166 in 1974, and the numeric country codes were first included in 1981. The country codes have been published as ISO 3166-1 since 1997, when ISO 3166 was expanded into three parts, with ISO 3166-2 defining codes for subdivisions and ISO 3166-3 defining codes for former countries

For UN, the list of countries or areas contains the names of countries or areas in alphabetical order, their three-digit numerical codes used for statistical processing purposes by the Statistics Division of the United Nations Secretariat, and their three-digit alphabetical codes assigned by the ISO.

## 2.2 DATA TYPES

Designing a map is not possible without knowing and understanding the nature of the data to be represented. Choices in the area of representation concern the graphic expression of the information (see Part 2). It is therefore essential to know how to characterize the data so as to process and represent it adequately. Below are a few practical elements to guide you.

### 2.2.1 Statistical Data Expresses Either a Quality or a Quantity

**Qualitative data** is not measurable; it involves names, acronyms, and codes. Qualitative attributes cannot be summed, and averages cannot be calculated. Qualitative data can be divided into two categories: ordinal qualitative data which can be classified in a given or chosen order, and nominal qualitative data which cannot be ordered. For instance, a hierarchical classification of European towns and cities – capital cities, regional capitals, secondary cities, etc. – is a form of ordinal qualitative data. Data on the official language of countries – French, German, Spanish, etc. – is nominal qualitative data that cannot be hierarchized.

**Quantitative data** is always numerical. By definition, the data is ordered, and average will have meaning. There are also two types of quantitative data. *Absolute quantitative data* expresses concrete quantities, and the sum has a meaning. There are two categories: absolute quantitative "stock" data corresponding to counts at instant t (e.g., the number of inhabitants on January 1) and absolute quantitative "flow" data (not to be confused with flow data expressing relationships between places), corresponding to counts over a period of time (e.g., the number of births in the year). Then, we have *relative quantitative data* derived from the calculation of a relationship between two values (for instance, the unemployment rate or population density). The sum of relative quantitative data, often expressed in percentages, has no meaning, and only the mean can be significant. By extension, composite numerical indicators can be combined with relative quantitative data mingling several simple types of data (e.g., indices).

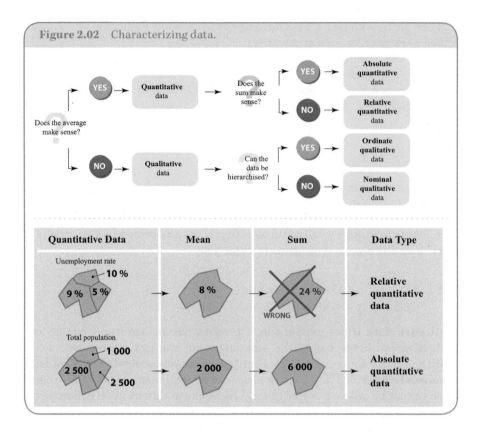

Figure 2.02    Characterizing data.

## 2.3 DATA PROCESSING

The statistical information contained in a data table cannot always be mapped directly. Most often, the information needs to be converted and collated, or reduced to render it intelligible (creation of indices, typologies, or classifications, etc.). The task is to order the information and to retain only what is useful for the cartographic representation. This work on the data, which is an integral part of cartographic construction, would merit a book in itself. We suggest readers refer to the bibliography to find the relevant references. Here, we discuss only what is essential in a book about cartography and map design. Complementary or interpretative information should be sought in the references provided.

In cartography, the graphic transcription of data cannot always be direct, since this could result in an unmanageable, illegible map. The data that needs always to be simplified is relative quantitative data (rates, indices, etc.) which need to be subdivided into classes of values. This procedure, known as **discretization**, is based on specific methods and characteristic values. The aim is to simplify the statistical series observed. To do this, there are several stages.

## 2.3.1 Summarizing

Reviewing and summarizing a statistical series is a way of becoming acquainted with it: what are the minimum and maximum values observed? How was the phenomenon measured? Is the value derived from a calculation? What calculation? Can the set of values be expressed by one or several characteristic values? Can spatial or statistical comparisons be made?

### 2.3.1.1 Position Parameters

Position parameters make it possible to sum up a statistical series in a single value. This can be formed by a specific value or by a value that is considered to be "central".

Specific values (minimum, maximum, or any other) are considered as being representative of a domain (for instance, the number of children required to renew the population), or they are fixed by a law or regulation (for instance, the occupation coefficient fixed by an urbanism document). Central values are calculated or determined from all the values in a series. There are three central values: the mean, the median, and the mode. The choice of the best central value depends on both the objective of the summary and the shape of the distribution.

**The mean value** indicated $\bar{x}$ (x bar) is the simplest statistical value expressing the magnitude of a statistical series. It is the sum of the values divided by the number of statistical units observed (or in cartography the number of geographical units). The mean is the gravitational center of the distribution: the sum of the deviations from the mean value is zero.

**The median** (indicated Q2) is the value that divides a statistical series into two parts comprising equal numbers. In other words, half of the values are above the median and the other half below. The median is the value that is nearest to all the values in the distribution.

**The mode** (or dominant value) is the most frequent value in a distribution. It is always calculated by scanning the set of values. A distribution can be unimodal (a single mode) or multimodal (several modes). In this case, it is usual to distinguish a main mode and one or several secondary modes.

> ### Definition
>
> #### Unit of Measurement and Order of Magnitude
>
> *The unit of measurement* is the unit serving to count or calculate the values in a series, for instance, the number of inhabitants per square kilometer for a population density, the percentage for urbanization rate, the hectare for a surface area, etc. Knowing the measurement unit enables the data on which one is working to be understood, and makes it possible to ascertain whether comparisons are possible with other data.
>
> *The order of magnitude* determines the variation or the extent of a series. It is provided by the minimum and maximum values in the series observed. It offers important information about the meaning of the data and the limitations for comparison with other data.

## 2.3.1.2 Dispersion Parameters

The notion of dispersion refers to the degree to which values in a distribution spread out or scatter one in relation to another or on either side of a central value. The assessment of dispersion is always linked to a central value. It indicates how far values in a distribution generally deviate or diverge from the reference central value

**The standard deviation** (indicated σ) is an absolute dispersion parameter linked to the mean. To calculate it, an intermediate calculation is required: the variance. The variance is a global measure of the variation of a set of numbers that are spread out from their average value. The standard deviation has a probabilistic meaning. Probability theory enables the estimation of the likelihood of a value to be distant from the mean by more than a certain number of standard deviations. Indeed, when a distribution is Gaussian, (also referred to as "normal" and characteristic of symmetrical distributions), the probability of finding values at a given distance from the mean is known. This property is very useful in cartography because it enables a rational subdivision of values in a distribution.

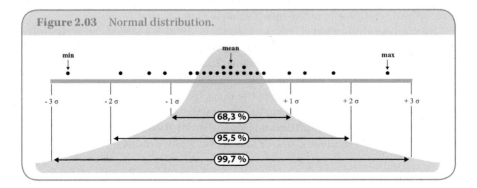

**Figure 2.03**    Normal distribution.

**The interquartile interval** is an absolute dispersion parameter linked to the median. It is defined as the extent of the distribution concentrating the central array of elements that differ, the least, from the mean. A chosen percentage of the highest and lowest values in the distribution are excluded. This parameter is linked to the notion of the quartile which defines the limits of a subdivision into classes with equal numbers. Thus, different intervals are described according to the desired subdivision, into 4 (quartiles), into 5 (quintiles), or into 10 (deciles), etc. For instance, the interquartile interval is the part of the distribution concentrating half of the elements for which the values are the least different from the median. Thus, 25% of the lowest values and 25% of the highest values are excluded from the distribution.

The comparison of the absolute dispersion parameters of a characteristic is only meaningful if these two characteristics are of the same order of magnitude. For instance, the comparison of two standard deviations can only be envisaged if the distributions have the same mean, failing which a structure effect can be favored. If they do not, the comparison is only possible by resorting to measures of relative dispersion.

A **relative dispersion parameter** is a measure of the relative deviation of values in a distribution in relation to a central value. It corresponds to an absolute dispersion parameter divided by a central value. One thus obtains a number with no dimension (the mean differences, i.e., differences in order of magnitude, have been removed). The most common relative dispersion values are the coefficient of variation (CV) = standard deviation/mean and the relative interquartile coefficient: 3rd quartile – 1st quartile)/median or Q3-Q1/Q2.

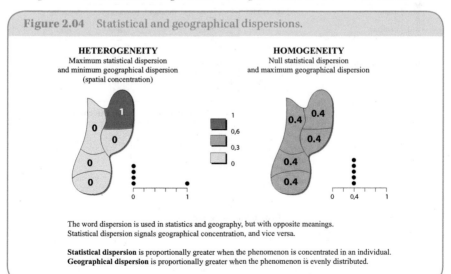

**Figure 2.04**    Statistical and geographical dispersions.

The word dispersion is used in statistics and geography, but with opposite meanings. Statistical dispersion signals geographical concentration, and vice versa.

**Statistical dispersion** is proportionally greater when the phenomenon is concentrated in an individual.
**Geographical dispersion** is proportionally greater when the phenomenon is evenly distributed.

## 2.3.2 Analyzing

The next stage is an understanding of the intrinsic characteristics of a distribution by exploring its shape and the dispersion of values. These two elements enable a subdivision into classes that are suited to the dispersion observed, and thus enable the right cartographic choices. The shape of the distribution can be determined from the observation of the distribution diagram or from the comparisons of the central values.

### 2.3.2.1 The Distribution Diagram

This enables the values in a series to be positioned along an axis that is oriented and graduated. The concentration of values on this axis reflects the concentration or dispersion of values.

If the values are clustered around a single concentration zone, the distribution is said to be unimodal. If the values are grouped around the mean value, the distribution is symmetrical. If they are concentrated around the low values, the distribution is asymmetrical or "skewed" to the left, and if they are concentrated around the high values, the distribution is "skewed" to the right.

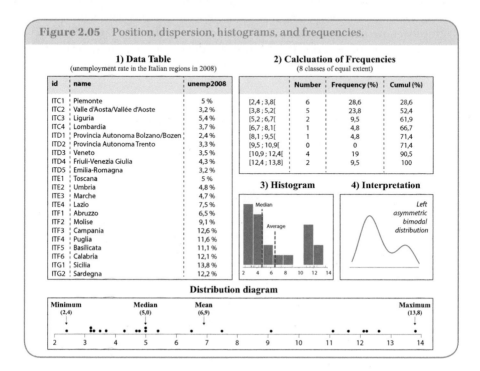

**Figure 2.05**   Position, dispersion, histograms, and frequencies.

If the values present two or several concentration zones, the distribution is skewed and bimodal or multimodal. In this case, the mean is not an appropriate means of summarizing, since it may well "fall" within a dispersion zone.

When the values to be observed are too numerous, their distribution can also be observed in aggregated manner. This operation involves a calculation of the frequency of values within classes of equal extent. The shape thus formed by the height of the different bars (histogram) reflects the distribution of the values.

### 2.3.2.2 Comparing Central Values

If the three central values (mode, median, and mean) are equivalent, the distribution is said to be symmetrical (or unscrewed). All the central values are relevant summaries.

If the central values differ widely one from another, the distribution is dissymmetrical or skewed. The mean is drawn towards the zone of dispersion; the mode is drawn towards the concentration of values. In this case, it is the median that provides the best compromise.

If the distribution presents no mode, the comparison can be made, on the same principle, comparing only the mean and the median values.

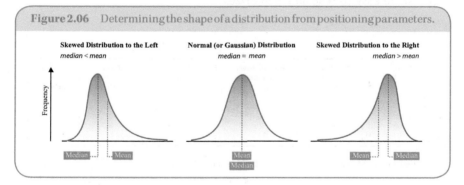

**Figure 2.06**   Determining the shape of a distribution from positioning parameters.

If the values present two or several zones of concentration, the distribution is skewed and bimodal or multimodal. Neither the mean nor the median is a suitable way to summarize the values, since they often fall within a dispersion zone. The principal mode, alongside the secondary mode, provides the best suited means of summarizing a bimodal distribution.

## 2.3.3 Determining Class Intervals

Discretization consists in subdividing a statistical series into classes of values. This operation needs to take account of the different characteristics of the distribution. It is part of the prior processing of information, the aim of which is a simplification of the information with a view to analyzing and/or representing it.

● **FOCUS: Principles of Data Classification**

*Principle no. 1:* the classes must be homogenous and distinct (no overlap and no break between classes), and the geographical objects in a given class should resemble one another more than they resemble objects in the other classes.

*Principle no. 2:* the number of classes is necessarily smaller than the number of observations, the merging of classes should correspond to the overall domain of variation for the characteristic under study, and the classes are ordered. The number of classes always depends on the number of statistical units observed, on the objective, and on the use to be made in the future (with or without a map).

*Principle no. 3:* the essential characteristics of the distribution should be preserved so as to lose as little information as possible. Three dimensions of the data series should be taken into account: the order of magnitude, the dispersion, and the shape of the distribution.

*Principle no. 4:* to facilitate the reading of the map, it is recommended to round off the boundary values for the classes, and if possible to use boundary values linked to relevant orders of magnitude. These boundary values should be easy to read and memorize.

**Figure 2.07**   How many classes?

| Territorial level (European Union) | Number of territorial units | Huntsberger method | Brooks-Carruthers method | Color perception threshold* | Maximun number of classes |
| --- | --- | --- | --- | --- | --- |
| NUTS 0 (Pays) | 28 | 6 | 7 | 8 | 6/7 |
| NUTS 1 (Länders) | 98 | 8 | 10 | 8 | 8 |
| NUTS 2 (Régions) | 272 | 9 | 12 | 8 | 8 |
| NUTS 3 (Départements) | 1 315 | 11 | 16 | 8 | 8 |

**Huntsberger Method**

$k = 1 + 3,3*\log_{10}(N)$

**Brooks-Carruthers Method**

$k = 5*\log_{10}(N)$

k: number of classes
N: number of territorial units

*(*) variable according to the spatial configuration*

There is no optimum classification. Each method will yield a different map, reflecting the actual distribution more or less efficiently. The aggregation of data into classes, in other words the reduction of the useful information, introduces an error or distortion in the perception of this distribution. In addition, the distribution pattern affects the choice of a method of classification.

This choice of a method of classification depends on the properties of the distribution and also on the cartographic objectives fixed. For instance, to put emphasis on very high values, the tendency will be to place them in a category apart so as to individualize them. Conversely, to represent a homogenous image of a territory, the tendency will be to choose a small number of classes.

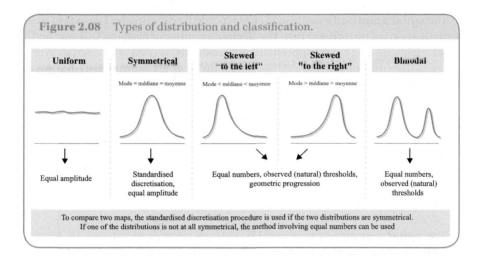

**Figure 2.08**    Types of distribution and classification.

| Uniform | Symmetrical | Skewed "to the left" | Skewed "to the right" | Bimodal |
|---------|-------------|----------------------|----------------------|---------|
| | Mode = médiane = moyenne | Mode < médiane < moyenne | Mode > médiane > moyenne | |
| Equal amplitude | Standardised discretisation, equal amplitude | Equal numbers, observed (natural) thresholds, geometric progression | | Equal numbers, observed (natural) thresholds |

To compare two maps, the standardised discretisation procedure is used if the two distributions are symmetrical. If one of the distributions is not at all symmetrical, the method involving equal numbers can be used

Ultimately, the classification procedure is guided by two sometimes contradictory objectives, which the cartographer will have to weigh up: preserving the characteristics of the statistical distribution so that the data will not be misleading, but at the same time allowing leeway to enable the delivery of an efficient cartographic message. When choosing a method of classification, the aim is thus at once to reflect the statistical series as accurately as possible, to give meaning to the different classes, to facilitate memorization, and to produce a clear message. A compromise between the statistics and the requirements of cartography is thus required. Choices must be made, but excessive manipulation should be avoided.

### 2.3.3.1 Classification Methods
The **Observed threshold** (also called natural threshold) method is conducted visually on the graphic representation of the distribution (distribution diagram), by identifying each "trough" or "hump" to define the class boundaries. This manual method enables a focus on discontinuities in a statistical series. The numbers in the different classes can be markedly unequal, and the subdivisions are subjective. As a result, this method cannot be used to compare two distributions.

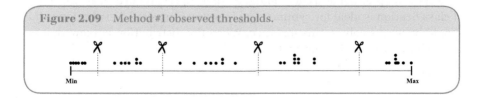

**Figure 2.09**    Method #1 observed thresholds.

The **equal amplitude** method is constructed by dividing up the extent of a statistical series (min/max) into the desired number of classes. This method, which yields simple, easily apprehended thresholds, is used with even or symmetrical distributions. It should be avoided for strongly skewed distributions. This method does not enable the comparison of several maps.

**Figure 2.10**    Method#2 equal amplitude.

min + a   min + 2a   min + 3a   min + 4a   min +5a

Min                                              Max

The amplitude of each class is defined by: $a = \dfrac{(\textbf{max - min})}{\textbf{k}}$

k : number of classes

The **equal numbers** (or quantile) method constructs classes in which there is the same number of statistical units. The classes formed in this way are known as quantiles. When there are 4 classes, the term used is quartiles (one quarter of the total number in each class), when there are 10 classes, they are known as deciles, and for 100 classes, they will be centiles. This classification method can be used with any type of distribution, and it enables the comparison of maps one to another.

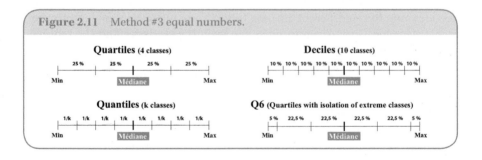

**Figure 2.11**    Method #3 equal numbers.

The **standardized classification** method uses significant values (mean and standard deviation). The value of the mean can appear, depending on the purpose of the map, either as a class boundary or as the center of a class. This type of classification is ideal for symmetrical (normal, Gaussian) distributions and should be used solely in this instance. When the distribution is skewed, it is preferable to use another method.

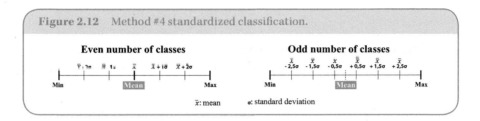

**Figure 2.12**    Method #4 standardized classification.

The **geometric progression** method is suited to highly skewed distributions. It consists in constructing classes whose extent increases (or decreases) with each class, which enables close follow-up of the statistical series. The task is to find a number (common ratio for growth) which by multiplication will give the amplitude of the class. This method assumes that the minimum is greater than 0.

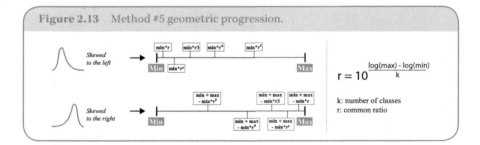

**Figure 2.13**    Method #5 geometric progression.

The Jenks (or Fisher) method is an automatic classification method based on the principle of resemblance or non-resemblance among individuals. The method functions via iteration. It groups individuals that most resemble one another and those that least resemble one another. In statistical terms, the method aims to minimize intra-class variance and to maximize inter-class variance. This method is on offer in most software and is a good "general" method adapted to all types of distribution. It should be noted, however, that it does not enable comparisons of maps one to another.

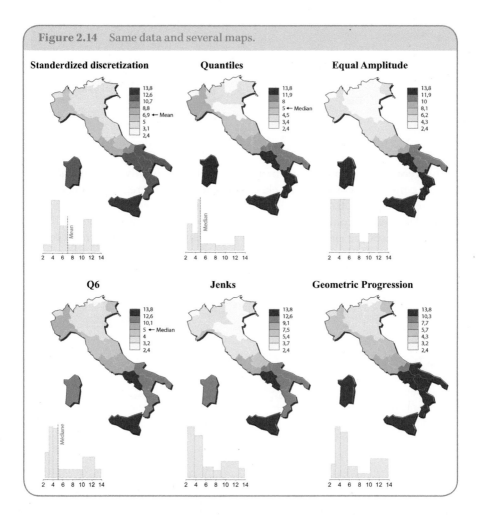

**Figure 2.14**    Same data and several maps.

● **FOCUS: Dividing Qualitative Data into Classes**

The subdivision of qualitative information into classes consists in a pre-defined, straightforward grouping of elements in smaller numbers so as to obtain a reasoned typology or classification. The operation is governed by the same principles as above for quantitative information. If the information relates to an order (ordinal qualitative data) the hierarchy of the informa-tion must be strictly complied with. If the information is mainly nominal, the information is grouped according to resemblance to form a typology. The formation of the classes is then specific to the objectives of the chosen simplification.

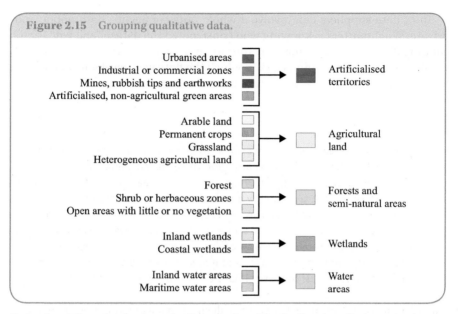

**Figure 2.15** Grouping qualitative data.

Corine Land Cover is a land-use database produced by the European Environment Agency. It covers 38 European states and provides a nomenclature according to three hierarchical levels (44 types in level 3, 15 in level 2, and 5 in level 1).

## 2.4 CAN DATA BE TRUSTED?

"Geographical data are not supplied by God, but by a given geographer who, not content to apprehend the data on a certain scale, has also chosen and ordered the elements in the dataset; another geographer studying the same region or addressing the same issue on another scale will probably come up with rather different data" (Lacoste, 1976).

Constructing statistical data and making it available was for a long time the exclusive domain of Nations. Data, a central component of national sovereignty, was often secret, hidden, and the various States only issued information in accordance with their strategic interests. But today, nation states are not the only official providers of data. At international level, numerous bodies collect or harmonize statistical data on different subjects of variable complexity. Thus, the OECD (Organisation for the Economic Co-operation and Development), the UN (United Nations), the IMF(International Monetary Fund), Eurostat, the World Bank, and even the CIA (Central Intelligence Agency) produce official data each year on different countries across the world. Whatever the scale, it is also possible to create one's own statistical data from surveys or measures in the field.

Thus, data suppliers can involve a whole range of agents, both public and private. What matters is knowing who constructed the dataset and who is circulating it. Thus is never innocent. Your analyses and results will be validated within the scope of the framework in which the data was developed and circulated. It is therefore always important to refer to the metadata.

### Definition

**Metadata** is data which describes other's data, data about the data. It is a marker that is applied to any type of resources, enabling its description: where does it come from, how was it created, and by whom? In fact, it ensures the traceability and the quality of the data (authenticating and assessing the data or the source). It also facilitates the search for information by describing its content, thus improving its referencing. It favors inter-operability by way of data sharing and exchange, improves data management and storage, and helps to manage and protect intellectual property rights. Statistical data cannot do without metadata.

A lot of data is used because it covers the desired space and the topic in exhaustive manner. For instance, data can answer a simple question such as "what are the unemployment trends in the different European region?" In this setting, it can seem of little importance what methods were used to collect the data and collate it, since the data required is available and official. Yet it is essential to remain critical with regard to the data so as to avoid misinterpretation or mistaken conclusions. This means that reflection is required beforehand on the intrinsic quality of the data used (accuracy, credibility, objectivity), on the context in which it is published (survey, census, GIS procedure, estimate, etc.) and the meaning (complex indicators, typologies, etc.). A sound approach requires the data supplier to be considered – who supplied the data? What other data does the supplier produce? For what purpose? What is the data intended to tell us?

It is also important to fully understand that any figure or value is the result of a construction. Statistical data used to produce maps is not a set of objective measures of a reality, but indicators enabling that reality to be approached, according to a specific construction and a certain aspect of that construction. What is behind the figures? Does a low employment rate in a country necessarily mean that access to jobs is easier than in a country with a high rate of unemployment? And even if the employment rates are similar, are the lives of individuals who are unemployed comparable? How precarious is employment? What access do workers have to healthcare? What are their rights? What is their place in society? And, in some countries, are there not people who are not accounted for by the official figures (on sick leave or after being struck off unemployment benefits)? In the words of Mark Twain (1835–1910), "Facts are stubborn things, but statistics are more pliable".

● **FOCUS: What Data Tell Us?**

Data is not innocent. Data contains a message. While relative quantitative data will tend to refer to a hierarchical and ordered view of geographical space, absolute data is in contrast closely linked to the notions of power and power balance.

For instance, while the GDP per capita of China ranks the country 120th in the world (according to the 2014 CIA World Factbook), the absolute value of its GDP ranks China 2nd among the main economic players internationally. Another example is if it is useful to know the number of armed forces per inhabitant to compare two countries, it is equally useful to know the absolute number of armed forces (or their military equipment), since in case of actual conflict, it will be the numbers that will matter.

The choice between absolute data and relative data thus involves two different views of the world: one that is more ordered, and other more conflicting. Thus, producing a map means that we need to take account of what the data has to tell us.

### Quiz

- What are the two parameters that must be taken into account when data is processed?
- What is a distribution diagram?
- What is the mean? The median? The standard deviation?
- How can a symmetrical statistical series be discretized? And an asymmetrical statistical series?
- What is metadata?

# Conclusion:
# Designing Maps

Any map starts with a blank sheet. The conception phase, requiring reflection, comes first. This step will condition or at least strongly orient the cartographic construction phase. This first phase requires thought, documentation, and hypotheses, so as to define a strategy.

The conception phase takes time. It is possible to create a reasonably good map in a few minutes, in particular with the support of mapmaking software. However an original idea can take years to mature before suddenly taking shape. An idea of this sort can appear quickly or unexpected, but somehow creativity must be worked on and fostered. The conception phase is also an exploratory phase in which the map designer will test different hypotheses and envisage the spatial configurations of the different pieces of geographical information. At this stage, it is possible to refer to geographical literature so as to appropriate different concepts, as well as to the literature of the arts or computer graphics, to work towards ways of putting ideas into visual form. To produce a map and to enjoy the process, the narrative needs to be captured. You need to test, try out, criticize, take the wrong path, and start again. It is a learning phase, in an attempt to create something that is both new and useful. Sometimes, a map is not the answer. If it is not useful, you should not hesitate to turn to other methods – graphs and tables, text, photographs, etc. Maps cannot do everything.

There are thus two stages, the conception and the actual production, pointing to two intellectual approaches. The first process (in green on the following diagram) tends to be governed by the data – its nature, how it is to be converted, how it is to be processed, its translation into graphic form, and the layout on a spatial medium. This conception process, which is exploratory, is the same as in the area of science. It makes it possible to understand how a phenomenon is organized and how it fits into space. The second process (in blue) reverses the reasoning by starting from a cartographic reflection and the idea of the message to be produced. From this cartographic intention or purpose will result the choice of the data, how it is processed, and how it is represented.

**Figure P1.C1**    Cartographic design paths.

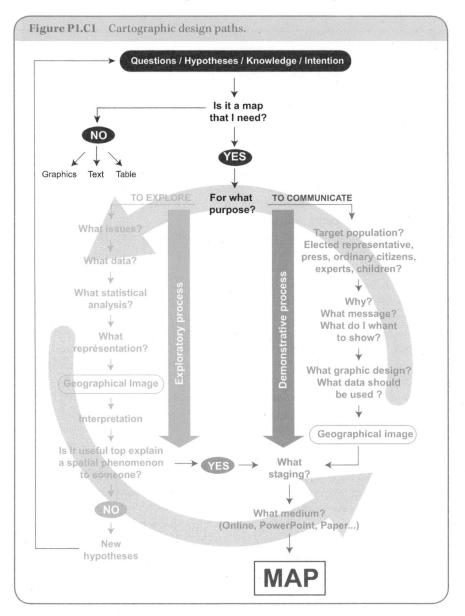

These two approaches to constructing a map are of course closely linked and overlapping. A scientific map can be "staged" in efficient and artistic manner; a communication map can be constructed with compete scientific thoroughness. Thus, the two approaches are complementary and function alongside, combining exploratory method with means of expression.

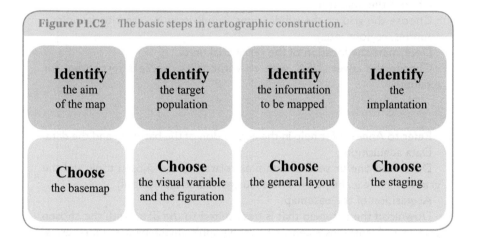

**Figure P1.C2**    The basic steps in cartographic construction.

**Identify**
the aim
of the map

**Identify**
the target
population

**Identify**
the information
to be mapped

**Identify**
the
implantation

**Choose**
the basemap

**Choose**
the visual variable
and the figuration

**Choose**
the general layout

**Choose**
the staging

● **FOCUS: MAPS or the Cartographic Method Design**

**M for message.** Designing a map means thinking what message is to be delivered. The message must be coherent, clear, intelligible, and as simple as possible so that it can be easily integrated. Cartographic images should sum up the geographical information in immediately perceptible form.

**A for artistic.** A map needs to be attractive. It should be aesthetic and well designed. Thus, a good map is one where the cartographer has paid attention to all the different elements provided and the layout, using color in a relevant manner.

**P for people.** The map should be designed specifically for a single purpose. Thus, the mapmaker needs to know whom he is addressing (a particular population) and adapt the processes used accordingly.

**S for scientific.** Never forget that a map is also a professional tool of the geographer. It is a scientific object complying with strict and precise rules. While a map is subjective, constructed according to a communication objective and from a particular point of view, it cannot be allowed to distort reality too far. Any map also needs to be reproducible and thus open to scientific contestation.

## THE MAP GAME PART 1: ACQUISITION AND PROCESSING OF THE DATA

### 1. Preliminary choices

• Choose the indicator.
• Choose the grid or the subdivisions (subdivision into territorial units – counties, regions, etc.).
• Determine the location of the territorial units.
• Choose the cover (USA, with or without its overseas territories, for example).

### 2. Acquisition

• Refer to Annex 2 for help in the acquisition of the basemap and the data.
• Data acquisition.
• Download one or several data files related to the chosen topic and the grid chosen (e.g., https://www.census.gov/datat/tables.html).
• Acquisition of the basemap.
• Download the basemap that is best-suited to the data, with the chosen grid (e.g., https://www.census.gov/geographies/mapping-files/time-series/geo/carto-boundary-file.html).
• Caution: ensure the compatibility of the basemap coding with the data (same number of territorial units, same codes for each unit). Do not forget to also download the metadata and store it.

### 3. Processing the basemap

• Edit (or "clean") the basemap: generalization of lines, elimination of non-useful elements, etc. according to the chosen grid and the format (image 800 pixels wide).
• If you use automatic generalization tools, take care to check the coherence manually, using graphics software.

### 4. Data processing

• Identify the nature of the indicator(s) to be represented.
• Should it be classified? Why?
• For a relative quantitative indicator (rate, index, etc.), determine the shape of the distribution and infer the best classification methods.
• Perform the chosen classification procedure (be sure to take account of the number of classes).

# Part 2

# Cartographical Language

*That's how I understood by the age of 12 that only cartography raises all the questions about the sea and the land. That is, the world, and our view of the world. You know what I mean?*

*Yeah, yeah, yeah. That's right. That's right.*

*I think I would have liked that, to be a cartographer.*

Jean-Claude Izzo, *Les Marins Perdus (Lost Seamen)*, 1997

## INTRODUCTION

How can we use an image to tell a story, denounce, demonstrate, or convince? How can a piece of geographical information be transcribed in graphic form? How can we design a spatial message? These questions sum up the challenge of cartographic language and its correct usage.

In the same way as a text can be described by way of the words it uses, its language register, and it grammatical rules, a map can be described by the cartographic language used, where the visual variables and graphic rules form its vocabulary and grammar (Zanin and Tremolo, 2003). This language, which is purely visual, is intended to simplify the reading of the message by constructing an image that favors spontaneous and immediate visual perception.

Like any language, the cartographer's language starts from a collection of elementary signs or graphic primitives: point, line, and area. They are generally localized on the basemap and refer to localities on the surface of the terrestrial space. Once these objects are materialized on the map, a thematic map will move on to a figuration phase, consisting in graphically transcribing or representing different pieces of geographical information.

Although it is governed and defined by strict rules, which will be detailed in this second part, cartographic language has enormous scope. Just as a writer does not need to add new letters to the alphabet nor to defy the rules of grammar to write a masterpiece, the cartographer can make a map "eloquent" by relying on construction rules that have been designed to be visually effective. These rules for the cartographer, far from constraining, offer wide scope for creation.

Several notions need to be defined so as to explore the challenges of cartography with the appropriate vocabulary:

**Localization** refers to the position of a place on the Earth's surface. It is a geographical notion.

**Positioning** refers to the graphic transcription of geographical objects on the plane of the map using points, lines, or polygons. This refers here to a cartographic concept. It relates to the basemap on which the map is constructed.

**Figuration** (or symbolization) is the graphic representation of data on the plane of the map. It involves simple graphic elements which contribute a figuration, a graphic sign, or symbol. Figurations may be independent from the positioning. For instance, proportional circles are point figurations (attached to a particular point on the map) that can concern both "point" objects (towns, meteorological stations, etc.) and "area" objects (regions, countries, watersheds, etc.).

A **visual variable** refers to the way in which graphic figures are made to vary so as to visually transcribe the variations of statistical data.

Jacques Bertin, in his work on graphic semiology (1967 and 1973), was the first to theorize what he called "retinal variables". His approach consisted in designing rules for the graphic transcription of information so that it is suited to the human eye, thus explaining the term. The use of retinal variables is explained by any graphic problem involving three or more components, when the two dimensions of the plane are already being used.

He describes six variables:

They are **size, value, color, shape, orientation, and texture (i.e., grain)**. A seventh variable is often added, that is, texture/structure. In many ways, this seventh variable can be considered as a particular variation of the visual variables shape and value. It combines the texture with a particular structure. It is however common in numerous manuals, in particular English-language publications, to have a semiology of only the six initial visual variables.

The set of rules governing the use of visual variables forms what is known as the graphic semiology. **Cartographic language** refers to the various graphic possibilities enabling the expression of cartographic messages.

● **FOCUS: Jacques Bertin**

Jacques Bertin is the father of graphic semiology. He was born in Maisons-Laffitte in 1918 and died in Paris in 2010, and was successively a CNRS research associate, head of the EPHE laboratory (Ecole pratique des Hautes Etudes), and director of studies in EHSS (Ecole des Hautes Etudes en Sciences Sociales). His approach was deliberately pragmatic. His aim was to facilitate life for map users and to make maps rapidly comprehensible to all by way of an efficient, universal graphic language, so that the map can be perceived in a "minimum instant of viewing" (instant minimum de vision).

Although graphic semiology, as conceptualized by Bertin, provided great details about organization and construction of graphic system for black-and-white maps (at the time color was not in use). Today, not only are maps increasingly viewed on screens, but also the vast majority is in color. To remain as close as possible to contemporary mapmaking practices, we will therefore devote particular detail to the aspects that are linked to the use of color, leaving aside the more subtle characteristics of black-and-white mapping. In fact, the aim of this handbook is not to give a precise description of the whole range of options provided by graphic semiology, as this has been done many times before. Our approach is set in an intentionally practical and pragmatic framework. We will devote the most attention to modes of representation that we consider to be genuinely useful and to be in line with contemporary usage in mapping.

Our choice is to approach the ways of using visual variables according to the nature of the data (the third component) that is to be transcribed by them. This way of considering graphic semiology provides a direct approach to the scope for visual perception and comprehension of relationships between different elements of data, and thus the spatial configuration they show up.

We can distinguish three types of relationship within datasets:

- **Relationships of differential type.** They express relationships of equivalence or difference, with no hierarchy or classification, and they will be transcribed by graphic associative and selective perceptions. This can concern, for instance, the nomenclature of objects or a typology. The visual variables that transcribe these relationships need to have a property of association (identical data are transcribed by way of graphic figures that are perceived as being the same) and a property of selection (different data are transcribed by graphic figures that are perceived as different).
- **Relationships of ordered type** express a hierarchy across the data (this is before that, this is more than that). The order can be intuitive in case of a qualitative characteristic (low, middling, high), or it

can express a numerical relationship (a density, a rate). This type of relationship is transcribed graphically by graphic figures that can be visually classified without any ambiguity.

- **Relationships of quantitative type** express raw (absolute) quantities that can be transcribed visually by way of proportionality.

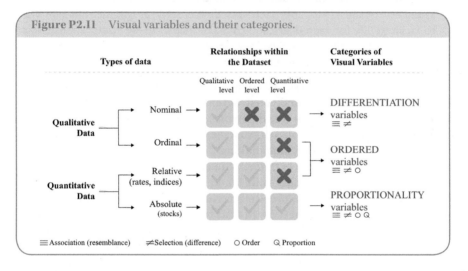

**Figure P2.11** Visual variables and their categories.

These three types of relationship enable the classification of visual variables in three categories: differentiation variables (Chapter 3), ordered variables (Chapter 4), and proportionality variables (Chapter 5).

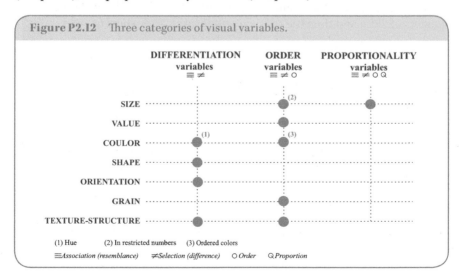

**Figure P2.12** Three categories of visual variables.

# Differential/Associative Visual Variables

## Objectives

- Getting to know the four visual variables used for differentiation and association
- Learning how to represent qualitative nominative data
- Learning to use all the dimensions of color
- Knowing how to choose the right color coding (RGB, CMYK, and HSL).

Nominal qualitative data should be "translated" visually by a graphic procedure that expresses differences. The procedure should enable grouping (association) and differentiation (selection) of the different geographical objects (this is like that, but different from this) without enabling the eye to classify them (no hierarchy). Several visual variables can be called on, of variable efficacy: color (hue), shape, texture/structure, and orientation.

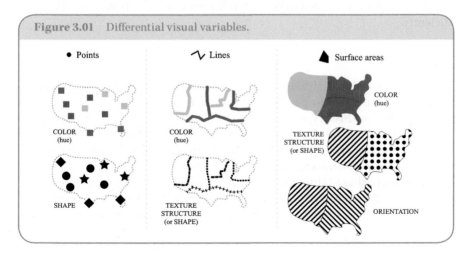

**Figure 3.01** Differential visual variables.

## 3.1 COLOR (HUE)

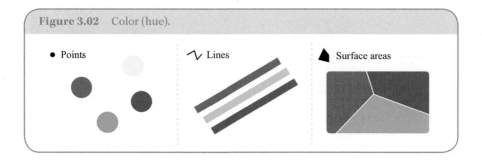

**Figure 3.02**    Color (hue).

Points

Lines

Surface areas

In the past, maps were often produced using shades of gray for reasons of printing costs, but today, with geographical maps increasingly intended for display on a screen, there is no need to do without color. Color as a visual variable facilitates the grouping and the differentiation of objects. Good harmony in the choice of colors also adds to the aesthetic effect and to attractiveness. However, we need to know how to choose colors appropriately. Beyond their mere physical properties enabling them to be recognized and differentiated by the human eye, colors carry symbolic and cultural meanings that also need to be considered. For point, line, and area figurations, color makes it possible to produce typologies. However, to avoid introducing an order between the different types, the intensity should be the same for all the colors used.

There are several principles governing the characterization and coding of colors. These principles need to be known if cartographic software is to be used efficiently.

### 3.1.1 HSL Coding

HSL (hue, saturation, lightness) coding is a model for the representation of colors that is intended to be close to natural perception by the human eye. The eye possesses three types of cone enabling the quantities of blue, red, and green contained in white light to be captured. The overall intensity of the light perceived and the proportions of the three primary colors enable the eye to perceive different colors according to three properties – hue, saturation, and lightness. According to this coding, for each hue it is possible to have the level of saturation vary (i.e., the purity of the color, whether it is bright or dull, as well as the lightness – the quantity of light in the color, which makes it lighter or darker).

### 3.1.2 RGB Coding

RGB (red, green, blue) coding is the colorimetric mode used for display on a computer screen. It is based on the principle of additive synthesis, each color being composed of a proportion of red, green, and blue. The proportions range from 0% to 100% (or from 0 to 255). Thus, the color blue is coded [R = 0%, G = 0%, B = 100%]. Yellow, which is a mixture of red and green in equal proportions, is coded [R = 100%, G = 100%, B = 0%].

Using the same principle, the **hexadecimal system** makes it possible to code colors according to a numbering system in base 16. It uses 16 symbols: 10 numbers (0 to 9) and 6 letters (A to F). According to the same logic as for the RGB coding, blue will be named simply #0000FF and yellow #FFFF00.

### 3.1.3 CMYK Coding

CMYK (cyan, magenta, yellow, key/black) coding is the colorimetric mode used in printing. It is based on the principle of subtractive synthesis. These colors correspond to the colors of the cartridges in a printer. Blacks defined as [C = 100%, M = 100%, J = 100%, K = 100%, K = 100%], [C = 100%, M = 100%, J = 100%, K = 0%], or [C = 0%, M = 0%, J = 0%, J = 0%, K = 100%] will therefore not have quite the same rendering on paper, and the quantities of inks used will not be the same. Thus, printers will often ask for an ink limit depending on the paper used.

Quadri-chromatic printing is however fairy rough-and-ready. If finer detail is required for the definition and the aspect of the different colors on the paper, more reliable solutions should be preferred.

● **FOCUS: Pantone – "The Power of Colors"**

In the 1960s, a US corporation, Pantone, set up a colors system enabling colors to be unambiguously identified. The idea was to provide printers with a very precise palette of colors, where each color had a specific code. Unfortunately, the system offered by Pantone is subject to intellectual property rights and is not available on free software.

### 3.1.4 Choosing the Right Colorimetric Mode

Since the RGB model and the CMYK model are based on very different techniques, it is not easy to switch from one to the other. Numerous algorithms are used in the software to enable this conversion, with highly variable results. In the conversion offered by the Office, a free software, for example, blue is logically coded [C = 100%, M = 100%, Y = 0%, K = 0%]. According to the

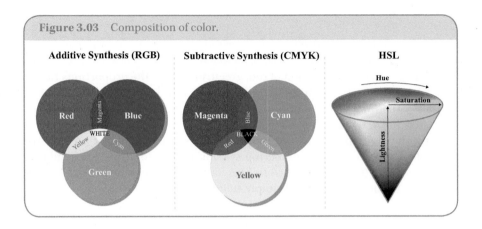

**Figure 3.03**    Composition of color.

conversion provided by Adobe Illustrator, this same blue is coded [C = 93%, M = 75%, Y = 0%, K = 0%]. To avoid problems when performing the conversion, the cartographer needs to define the colorimetric mode of his or her document from the outset: RGB if the map is to be displayed on a screen, and CMYK if it is to be printed.

## 3.1.5 Harmony and Perception of Colors

In cartography, as in dress, not all colors "go together". While certain colors cannot be combined for purely aesthetic reasons, the handling of colors should above all take account of the efficient perception of the different information that this visual variable transcribes in graphic manner. The operation requires skill. Colors are perceived differently according to context, so that using them is not an easy task.

The figure below comprises two small squares inside two larger rectangles, in shades of gray and in color. The two small squares (gray or green) are identical, and yet those on the left appear darker than those on the right. In fact, when the square is on a light ground, the eye tends to darken the square in order to see it better, while on a dark ground, the eye tends to lighten it.

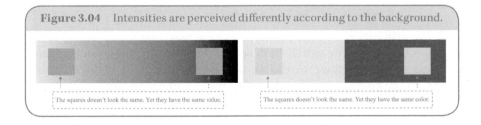

**Figure 3.04**    Intensities are perceived differently according to the background.

The squares doesn't look the same. Yet they have the same value.

The squares doesn't look the same. Yet they have the same color.

In cartography, color is a major variable that is immediately and intensely perceived, so that it has strong differentiation potential. The use of color is appreciated by both mapmaker and map user. It provides enormous scope for visual variation and helps make a map pleasant to look at.

## 3.1.6 The Symbolic and Cultural Dimensions of Color

Nevertheless, in visual communication the use of color is complex, and cartography is no exception. Colors, which possess intrinsic physical properties (each color corresponds to a specific wavelength), are also subject to strong, pervasive, psychological connotations. Thus, the task of giving them meaning and using them appropriately is relatively subtle. While there are objective criteria – for instance, warm colors have a longer wavelength than cold colors – each human being is educated from childhood to interpret colors in a particular way. Thus in Western cultures, warm colors suggest either warmth or danger, while cold colors suggest peace and wide icy spaces. In fact, a given color can call on opposite notions. Allocating a color to a geographical phenomenon is neither easy nor innocent.

● **FOCUS: The Semantics of Color**

**Blue** is a transcultural color, the color of the sky on any point of the globe. It is suggestive of consensus and wisdom. It was neglected in Antiquity because it was difficult to obtain, and it was the color of the Barbarians and of foreigners. Today blue is not suggestive of anything hostile, and it has lost its symbolic value and has become a discreet color.

**Green** expresses fertility and paganism in Europe. For the Moslems, it symbolizes Islam and mourning in Asia. Green is considered to be evil or bad luck in the world of entertainment. It at once suggests cleanliness and nature, and also deceit and hypocrisy. It's the Irish national color.

**Red** symbolizes fire, blood, and love, and also hell. Red draws the eye and is the color of danger (in the natural world, what is red is often poisonous or venomous). In politics, red is also communism. On numerous historical maps, the Soviet Union is presented in red. It is the color of pride, power, and ambition; it is insolent, violent, and associated with sin and crime. Yet in India, it is a symbol of purity.

**Yellow** is the color of infamy, the foreigner, the traitor, and the one that cannot be trusted. Judas's robe was yellow. But in Antiquity, the color was prized. In China, yellow suggests wealth and wisdom, and was long the reserve of emperors. In the Philippines, yellow symbolizes peace and resistance.

**Brown** is the color of mourning in India, the color of the Nazis in Europe, and the color of ceremonies for the Australian Aborigines.

**Purple** signifies dignity, royalty, elegance, and wealth. It is also associated with death and the crucifixion (especially in Europe), and with prostitution (in the near East).

White is the color of purity and innocence, and the light of God. But it is associated with mourning in India and Japan.

**Black** recalls fascism, anarchism, and other extremist movements with negative connotations. Black is also associated with mourning, but not in Asia where it is linked to death (while mourning is generally in white). Black is the opposite of white.

Gray, between white and black, is dull and unobtrusive; it expresses tranquility or even sophistication. It is at the same time quite easy to use because it can be associated with any other color.

Beyond the cultural meanings of the different colors, maps often use just the symbolic value of the colors observed in the natural world. When colors conform to what is directly observable in the environment, perception and comprehension can be facilitated. Thus, blue is widely used for seas and lakes, green for vegetation and forest, brown for land or for mountain areas, while red or purple can suggest human presence (as these colors are less common in the natural world). White signals the absence of information (recalling the large, uninhabited spaces of the Polar Regions). However, the use of this symbolism is optional in theme mapping. Each mapmaker is free to establish his or her own graphic system according to the purpose of the map and editorial constraints.

### 3.1.7 Colors Give Meaning to Maps

The use of color is an efficient means of giving a map more meaning, but this means knowing who is addressing. It is indeed easy to misuse color, with counter-productive effect. This is particularly true when maps are produced in a supranational context, as in the setting of Europe. Green for instance does not symbolize the same thing in France (nature conservation) as neither in North Africa (Islam) nor in Ireland where it is the national color. Thus, the use of this color will be perceived differently according to the nationality, the cultural origin, or the religion of the map user. Yet when used correctly, color is a powerful variable in mapping. Beware however of clichés – are we sure we want girls in pink and boys in blue on our geography maps?

Probably not....

● **FOCUS: The Impact of Colors**

The map shown below is taken from *The Clash of Civilizations* by Samuel Huntington published in 1997. In this essay, which aroused considerable debate, Samuel Huntingdon draws the portrait of a world that is fragmented into blocks of civilization where the cornerstone is religion. The original map in the book is in black and white. The colors chosen here are those used in the French television program "Le dessous des cartes" (Maps Underlines) in July 2002, for the same subject. In this program, which was nevertheless critical of Huntington's map, the cartographic language used is highly evocative: red for the communists, green for Islam, and blue for the West. Proximities in color reflect proximities between civilizations (blue/violet to suggest proximity between the West, the Orthodox world, and Latin America): a caricature of a map, for a theory that was equally caricatural.

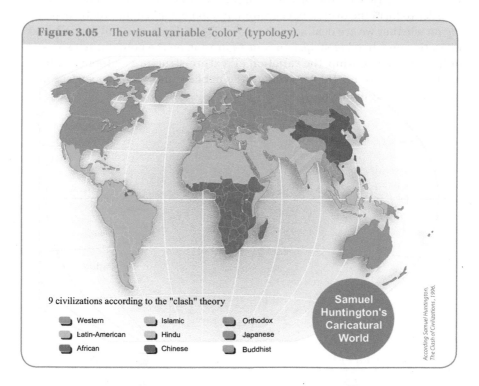

**Figure 3.05**  The visual variable "color" (typology).

9 civilizations according to the "clash" theory

| | | |
|---|---|---|
| ● Western | ● Islamic | ● Orthodox |
| ● Latin-American | ● Hindu | ● Japanese |
| ● African | ● Chinese | ● Buddhist |

Samuel Huntington's Caricatural World

According to Samuel Huntington, *The Clash of Civilizations*, 1996.

## 3.2 TEXTURE/STRUCTURE (WITHOUT HIERARCHY)

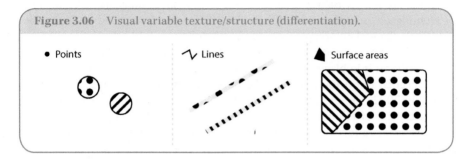

**Figure 3.06**   Visual variable texture/structure (differentiation).

The visual variable "texture-structure" is the term to refer to a pattern. It consists in varying the distribution and positioning of a graphic element in order to form the overall background. It can be applied differently depending on whether we are dealing with a point figuration on a small disc, a linear figuration on a band or strip, or a zone figuration on the polygon. To express a differential relationship, the variation of texture-structure (without any hierarchy) needs to retain the same quantities of black and white per surface unit in homogeneous manner.

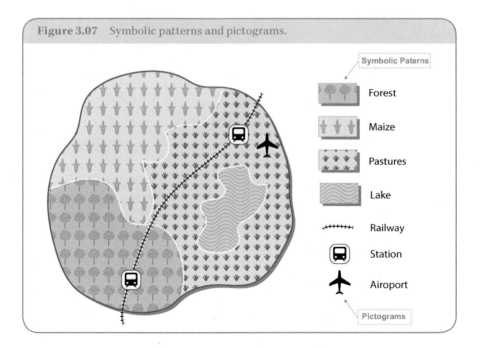

**Figure 3.07**   Symbolic patterns and pictograms.

In practice, this means making a graphic element vary (a texture, for instance, using hatching, dots, crosses, or any evocative motif) at defined intervals (providing the structure of the ground). If the ground is made up of evocative signs, they are referred to as "symbolic patterns" (i.e., repetitive elements) rather than texture-structure. This visual variable is mainly used for areal objects. It is not well suited to objects with a small area nor to small objects (dots or lines).

## 3.3 ORIENTATION

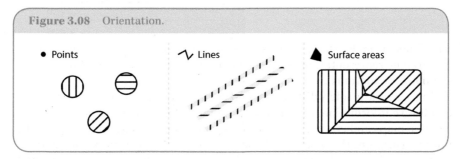

**Figure 3.08**    Orientation.

The visual variable "orientation" consists in varying the angle of a figurative element (hatching, crosses, etc.) so as to represent a qualitative difference. Although it is not very effective, this visual variable is mainly used on elements with a large surface area and using only four orientations: horizontal, vertical, and 45° in either direction. This variable, which was very useful when color was not easy to manage and maps were made manually, is no longer very effective, especially on a computer screen where the lines making up the texture are not easy to see.

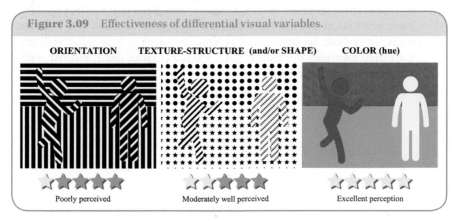

**Figure 3.09**    Effectiveness of differential visual variables.

## 3.4 SHAPE

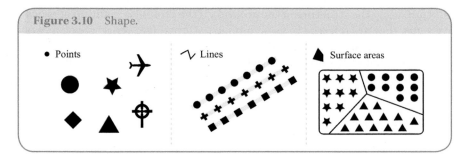

Figure 3.10   Shape.

Points    Lines    Surface areas

The visual variable "shape" consists in having the geometrical contours vary in a graphic representation. Shapes can be geometrical (squares, stars, etc.) or symbolic (pictograms).

This variable is solely differential and enables the representation of nominal qualitative information. In punctual symbolization, it enables the creation of location maps and is widely used for topographic or chorographic maps. With linear figuration, the visual variable "shape" amounts to vary the texture-structure. Whatever the figuration or the positioning, this variable is restricted in its variation. The eye does not efficiently perceive more than three or four different shapes. The variable should therefore be used sparingly.

▌ Definition

**The chorographic map:** chorography is a term derived from the Latin word *chorographia* and the Greek word *khôrographia*. It is formed from *khôra* (country, territory) and *graphien* (write). It refers to the description or representation of a country, a region, or a given space. Chorographic maps are thus at the interface between topographic maps and thematic maps. They describe both natural elements (relief, hydrography, etc.) and anthropic elements (statistical data, infrastructures, etc.). Tourist maps are generally chorographic maps.

**Figure 3.11**   Example of a chorographic map: Guadeloupe.

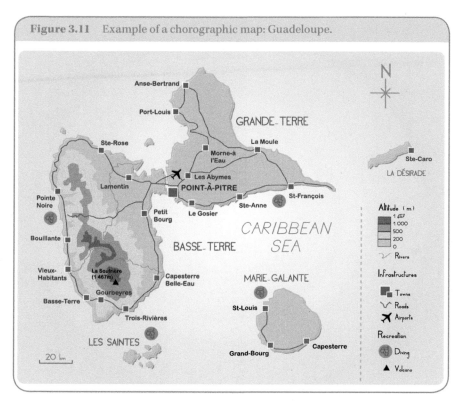

*This chorographic map of Guadeloupe describes both natural elements (relief, water courses) and human developments (roads, towns, airports). Leisure venues are also indicated as on most tourist maps (diving, volcano, etc.)*

## Quiz

- What do we mean by the cultural dimension of color?
- What do we mean by the symbolic dimension of color?
- To post a map on a website, what color coding should we choose?
- What is a chorographic map?
- What is meant by the word pattern?

# Chapter 4

# Ordered Visual Variables

**Objectives**

* Getting to know the visual variables that can express order
* Knowing how to represent ordinal qualitative data
* Knowing how to represent relative quantitative data.

When statistical data can be classified and hierarchized, it should be transcribed graphically using a visual variable that is able to express order. These variables can be applied to ordered qualitative data and to relative quantitative data.

Several visual variables can be used: **size, value** (or the intensity of the color), **ordered colors (color lightness), grain, and texture-structure**. These visual variables can only transcribe at best seven or eight different, ordered values. We therefore need to discretize the initial information, i.e., build homogeneous classes of values, so as to render the information intelligible (see Chapter 2).

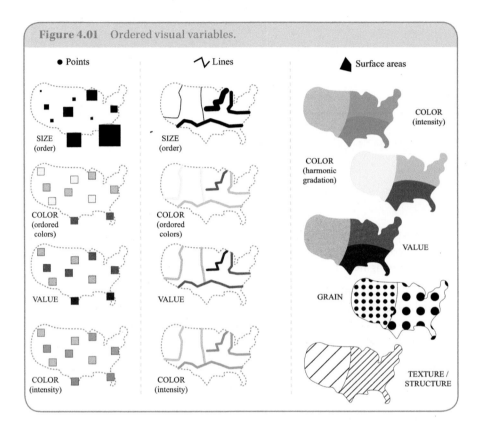

Figure 4.01    Ordered visual variables.

## 4.1  SIZE (ORDER)

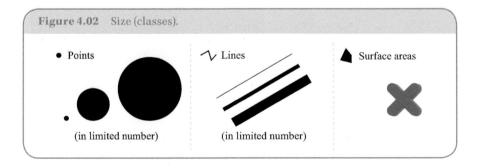

Figure 4.02    Size (classes).

The visual variable "size" is a procedure consisting in varying the height, the surface area, the thickness, or the volume (in 3D) of a geographical object. Size, often used to transcribe absolute quantitative data, can also be used to process discrete data (qualitative or quantitative). In this case, the visual variable "size" is constructed by creating classes of surface area, thickness, or volume reflecting a hierarchy between the geographical objects observed.

This variable can be used, in punctual figuration, to represent, for instance, types of towns or cities according to size (small towns, medium-sized towns, capital cities, conurbations), or in linear figuration to describe a road network (minor roads, main roads, motorways). This variable is not used in areal figuration. Care should be taken here: the visual variable "size" should not give an impression of proportionality (see Chapter 6). To avoid this, the number of different sizes on the map should be small and the sizes should be clearly different one from the other. Thus, it is not possible to use the visual variable "size" to transcribe relative quantitative information without prior discretization into a restricted number of classes.

## 4.2 THE VALUE AND INTENSITY OF COLOR

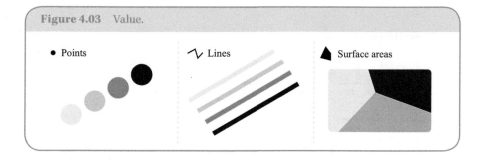

Figure 4.03    Value.

The visual variable "value" is the relationship between the quantities of black and white perceived on a given surface area. This means a gradation across levels of gray, from light gray to black, white being generally used to indicate the absence of data on certain geographical objects for the phenomenon observed.

This procedure can also be applied to a color by changing the relative degree of black or white mixed with a given hue (lightness) or by varying the purity (saturation). Here, we refer to shades of color.

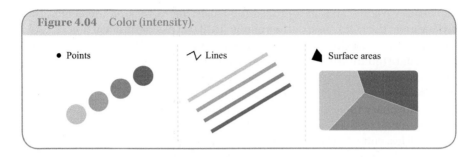

Figure 4.04    Color (intensity).

The human eye is able to rapidly classify dark and light elements in an image and to classify them from the lightest to the darkest, or vice versa. This visual variable (value using shades of gray or color gradient) makes it easy to represent an order, enabling the rapid visualization of spatial structures. This variable is applied to ordinal qualitative or relative quantitative data (after discretization). Its use on areal objects (polygons) enables choropleth maps. Their efficiency in delivering a message is greater when the number of shades used is fairly small, fewer than seven, and when the thresholds are well defined, so as to take account of the limits imposed by the physical performance of the eye.

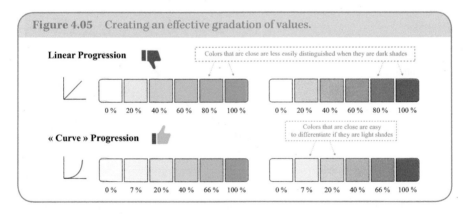

Figure 4.05    Creating an effective gradation of values.

## Definition

**The choropleth map:** the term "choropleth" was invented in 1938 by the American geographer John K. Wright. It is derived from the word *plethos* meaning quantity and the word *khôra* meaning space (country, territory). In geography, a choropleth map refers to a cartographic representation formed of ordered colored areas (or shades of gray) on some sort of grid (i.e., administrative division). This type of cartographic representation is used with relative quantitative data and ordered qualitative data.

## 4.3 ORDERED COLORS

Another way to use color to represent an order is to use the properties of the visible spectrum of light that is to say all the monochrome components of light that are visible to the human eye. In fact, each color corresponds to a specific wavelength which can be classified in an objective manner. Thus, the physical order of colors is measurable and universal. It ranges from dark blue to dark reddish-brown, through green, yellow, orange, and red. These are the colors of the rainbow, which enable us to establish meaningful color palettes.

The use of color palettes based on wavelengths thus enables the graphic translation of a relationship of order across values using their position in the visible spectrum.

But be careful! There is no direct relation between the wavelength colors and the perception of color orders on a map. For example, blue and green colors are difficult to order: which one came before the other?

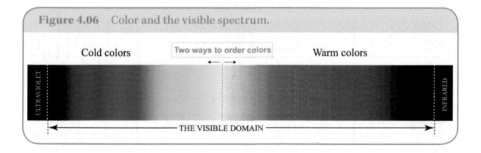

**Figure 4.06**    Color and the visible spectrum.

Nevertheless, the distinction between "warm" and "cold" colors is very useful to show up contrasts between high and low values in a statistical series, knowing that the contrast is always in reference to a meaningful value (zero, mean, median, etc.). In this case, the gradation is said to be balanced or two-sided.

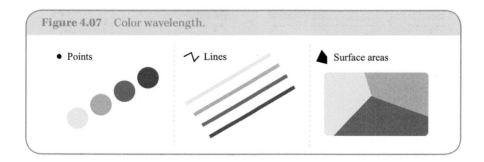

**Figure 4.07**    Color wavelength.

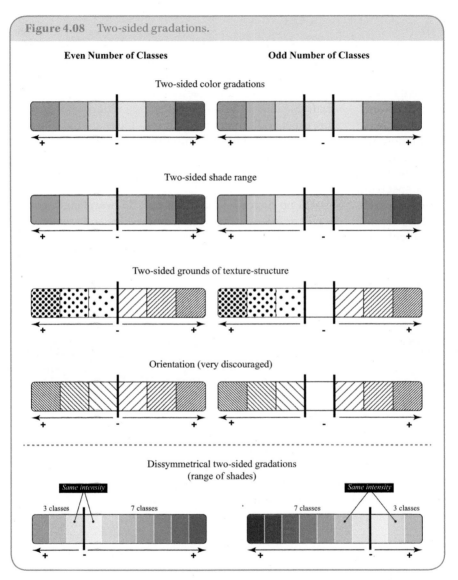

**Figure 4.08**   Two-sided gradations.

To design a map with a two-sided gradation, the choice of the colors is crucial. While the contrast between cold and warm colors wavelength gradation) is the most commonly used approach, because it reflects the contrast between geographical objects in the most direct manner, the use of two ranges of shades is also possible. When the two-sided gradation is dissymmetrical (more classes to one side of the reference value than to the other), the choice of the color intensity is the determining factor in showing the organization of values across a map. Very few software programs offer palettes that are directly available.

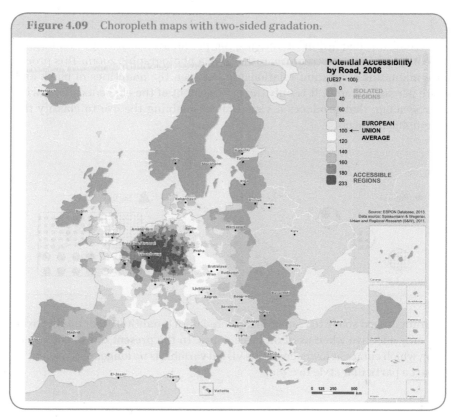

**Figure 4.09**    Choropleth maps with two-sided gradation.

*This map represents the potential accessibility by road at any point in Europe. By using diverging scheme, the perception of the central area is reinforced by the opposition between isolated regions (in green) and accessible regions (in red).*

● **FOCUS: The Origins of Thematic Mapping**

The first theme map in the modern era is attributed to a Frenchman, Charles Dupin (1784–1873). Charles Dupin was a scientist by training and was interested in social and economic statistics. In 1826, in the course of a lecture he presented the first figurative map of popular education. This map, using shades of gray for the educational levels of the different French departments (administrative areas), was the first choropleth map in history. The use of the grayscale is not insignificant on this map. White represents the knowledge (Age of Enlightenment), and the black refers to ignorance (obscurantism). While at the time these maps were referred to as "figurative", "special", or "applied", it is only since the 1960s that they have been known as thematic maps. ➲ *https://frama.link/cdupin*

## 4.4 GRAIN

The visual variable "grain" is the graphic procedure consisting in enlarging or reducing a texture-structure, rather like a photographic zoom. This procedure maintains a constant relationship between the quantities of black and white per surface unit. It is only the enlargement of the elements, along with their spacing, that produces a visual effect enabling the eye to classify the geographical objects.

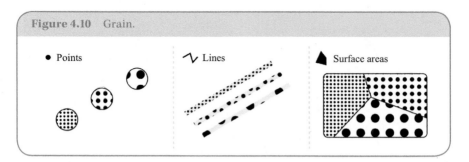

**Figure 4.10**   Grain.

Grain is used solely to transcribe ordered data (quantitative or qualitative). It is only effective when used on large zones. In the present day of the use of color, which comes at lower cost, this visual variable is no longer used except to impart a particular style to a map.

## 4.5 TEXTURE/STRUCTURE (ORDER)

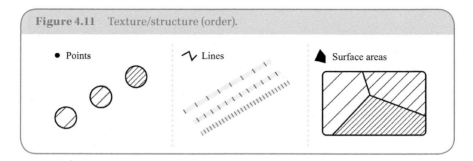

**Figure 4.11**   Texture/structure (order).

While texture-structure enables objects to be distinguished one from the other, this visual variable can also be used to represent an ordered relationship. To achieve this, the order is constructed using variations in spacing and thickness between the elements making up the ground (dots, hatching, etc.) by introducing a variation in the proportions of black and white per surface unit. In the same manner as for the visual variable "value", this procedure thus consists in constructing grounds of varying density so as to produce elements that are light and dark. This procedure is mainly used for areal units.

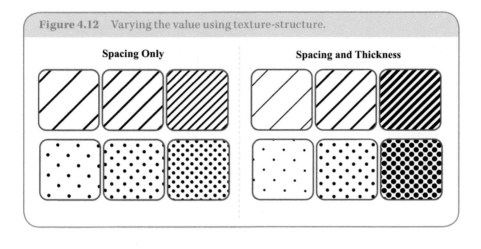

**Figure 4.12**    Varying the value using texture-structure.

## Quiz

- What is a choropleth map?
- What is a two-sided gradation?
- Why do we need to restrict the number of shades of color on a map?
- What is meant by shades of colors?

# Chapter 5

# The Visual Variables
# Expressing Proportionality

## Objectives

- Knowing how to use the visual variable "size"
- Knowing how to represent absolute quantitative data
- Knowing the fundamental rules for constructing the visual variable "size".

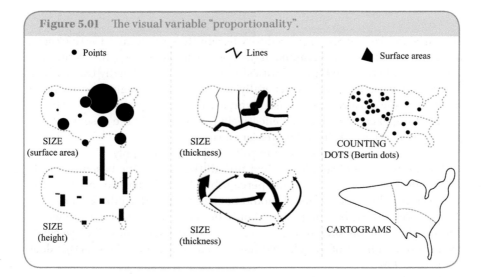

**Figure 5.01**    The visual variable "proportionality".

Points

Lines

Surface areas

SIZE
(surface area)

SIZE
(thickness)

COUNTING
DOTS (Bertin dots)

SIZE
(height)

SIZE
(thickness)

CARTOGRAMS

Absolute quantitative data is ordered by nature. However, the main characteristic that we are seeking to transcribe when we use this sort of data is the proportionality relationship inherent in the nature of the data. For this type of data, it is possible to sum the values of a phenomenon for two geographical units so that they can be aggregated. This property, which is specific to absolute quantitative data, needs to be transcribed graphically. The only visual variable able to achieve this is the visual variable "size". It can be noted, finally, that unlike relative quantitative data, absolute quantitative data should never be discretized.

## 5.1 SIZE (PROPORTIONALITY)

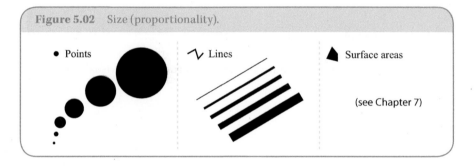

**Figure 5.02**    Size (proportionality).

The variable "size" involves a procedure whereby the surface area, the thickness, or the volume (in 3D) of a graphic figuration are made to vary. This is the only visual variable that enables quantities to be directly graphically transcribed. The visual variable "size" is generally applied to a straightforward point figuration. The most commonly used symbols are the square and the circle. The bar (rectangle) can also be used effectively, but it requires the basemap to be put in perspective (3D effect) for greater clarity.

For reasons of readability, certain rules and strategies apply.

The largest symbol should not "eat up" all the area of the basemap, while the smallest symbol should remain perceptible. A careful balance between the smallest and the largest symbol should be maintained. If symbols overlap for lack of room on the basemap, they should be outlined, for instance, in white. The smaller symbols should be positioned over the larger ones. The eye easily restores the segment of a circle that has been amputated, so long as this does not exceed 45%.

When the figuration on the map is of the zone type, we need to determine an "anchorage" point for the symbol. In automatic mode, this can be assimilated to the center of gravity of each territorial unit. It is also possible to choose one or other specific location that best sums up the territory in question.

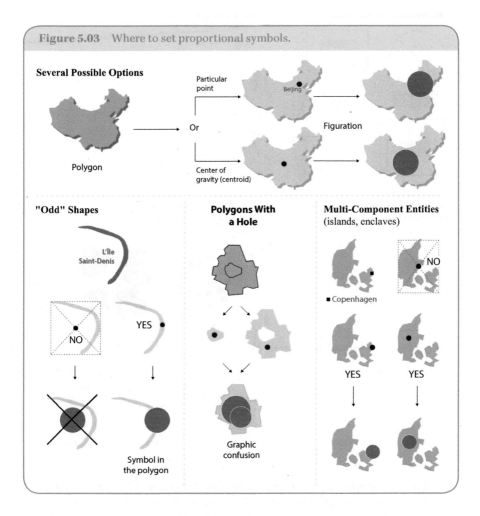

**Figure 5.03**  Where to set proportional symbols.

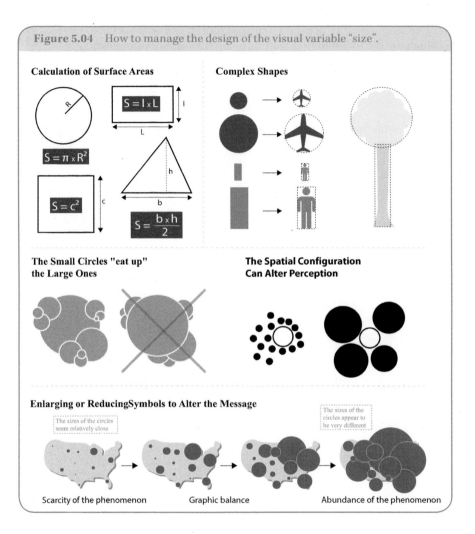

**Figure 5.04**    How to manage the design of the visual variable "size".

**Calculation of Surface Areas**

$S = l \times L$

$S = \pi \times R^2$

$S = c^2$

$S = \dfrac{b \times h}{2}$

**Complex Shapes**

**The Small Circles "eat up" the Large Ones**

**The Spatial Configuration Can Alter Perception**

**Enlarging or ReducingSymbols to Alter the Message**

The sizes of the circles seem relatively close

The sizes of the circles appear to be very different

Scarcity of the phenomenon          Graphic balance          Abundance of the phenomenon

In linear figuration, size, expressed by the thickness of the lines, makes it possible to produce maps of networks, flows, or discontinuities (see Chapter 7). In areal figuration, the use of the visual variable "size" consists in distorting the basemap itself so as to associate the surface area of each territorial unit with a quantitative variable, in which case we refer to a cartogram or anamorphosis (Chapter 7).

While the use of proportional symbols is quite easy to perform on cartographic software, the way they are used can be parameterized. Starting from the same statistical data, it is possible to produce several different maps. The determination of the optimal size to achieve graphic balance can be performed automatically. But this choice is not an end in itself. If a cartographer wishes to stress the general lack of magnitude of a phenomenon, or conversely its magnitude, he or she can very well choose to increase or decrease the size of all the symbols. From a visual viewpoint, the larger the symbols, the more obvious the differences between the different elements. The graphic transcription of data using the visual variable "size" is therefore not a purely automatic procedure.

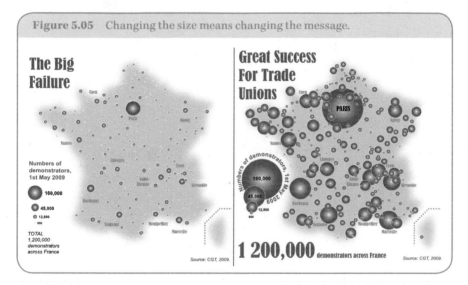

**Figure 5.05**    Changing the size means changing the message.

*This map shows the distribution and the scale of union demonstrations on May 1, 2009, in France. On this image, the basemap is toned down, and the surface areas of the circles (point figurations) represent the numbers of demonstrators in the main French cities (punctual symbolization). The cities with the largest numbers of demonstrators are explicitly named on the map. When the size of the circles is changed, the message on the map radically changes. Beware of manipulation.*

## 5.2 VARYING HEIGHT AND VOLUME

The new possibilities for 3D representation provided by cartographic software or GIS today enable the use of the visual variable "size" in many ways: extrusion, height, volume, etc.

In three dimensions, when applied to a point object, the visual variable "size" is most of the time represented as bars of varying height. On this type of map, although the comparison of heights that are not uniformly distributed across the space of the map is not easy, there is the advantage of producing an image that is relatively appealing. Other graphic constructions are also possible – spheres, cubes, etc. In this case, the statistical data will be represented by the volume of the object in 3D.

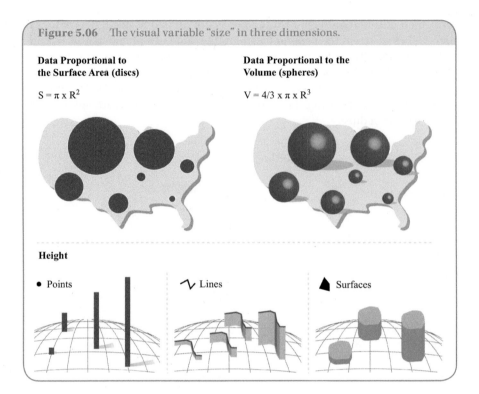

**Figure 5.06**    The visual variable "size" in three dimensions.

**Data Proportional to the Surface Area (discs)**

$S = \pi \times R^2$

**Data Proportional to the Volume (spheres)**

$V = 4/3 \times \pi \times R^3$

**Height**

• Points        ∿ Lines        ▲ Surfaces

For lines and polygons, objects that already have a geometrical shape on the map; the visual variable "size" in three dimensions consists in "extruding" the objects. For lines, the variation of size forms ribbons of varying height in three-dimensional spaces. When height is applied to polygons, they are expanded upwards in proportion to the value to be depicted. An "extruded" map, not very effective when there are too many territorial units, has the advantage in certain cases of producing a synthetic image enabling the viewer to perceive the order of magnitude of a spatial phenomenon in a few seconds.

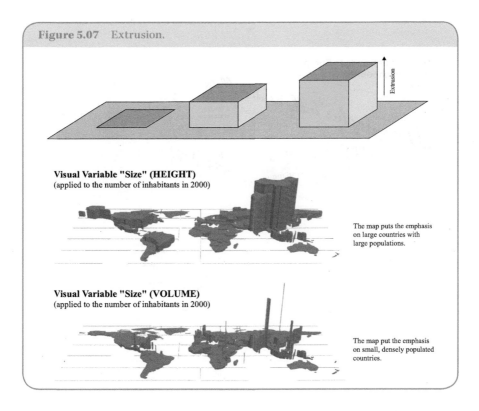

**Figure 5.07**  Extrusion.

**Visual Variable "Size" (HEIGHT)**
(applied to the number of inhabitants in 2000)

The map puts the emphasis on large countries with large populations.

**Visual Variable "Size" (VOLUME)**
(applied to the number of inhabitants in 2000)

The map put the emphasis on small, densely populated countries.

It can also be noted that application of this variable is open to discussion when the territorial units are very heterogeneous in size. Indeed in addition to the height of objects, the eye will tend to also perceive their volume. So in order to not to mislead the user, a methodological note should be provided to detail the procedure which has been used.

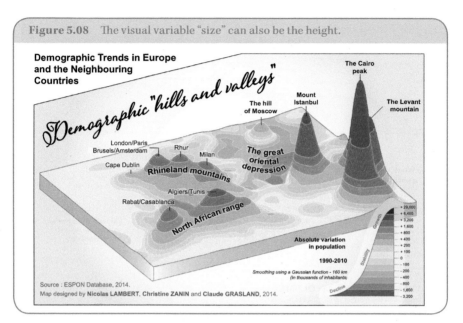

**Figure 5.08**    The visual variable "size" can also be the height.

*This three-dimensional map creates an imaginary landscape, with mountains and valleys, plains, and plateaus. Each peak describes an increase in population and each hollow a decline in population. This gives us a picture of our recent demographic history. Three main spaces can be seen. To the north-east is the large oriental depression overlooked by the "hill" of Moscow. Presenting as a hollow on the map, this depression, which follows the line of the former iron curtain to the east, is characterized by the loss of 4.5 million inhabitants in under 20 years for Russia alone. To the south-east is a region with three peaks culminating on the Istanbul Mountain, the Levant Mountain, and the Cairo peak, which dominate the whole region. This region is characterized by strong population growth over the 1990–2010 period. Finally, to the west, two mountain ranges of similar size face one another. The Rhineland range, on which are perched the cities of Milan, Paris, or London, and the North-African range dominated by the cities of Rabat, Casablanca, and Tunis.*

## 5.3 OTHER APPROACHES: "BERTIN'S" DOTS AND DOT DENSITY MAPS

In case of an aerial figuration (basemap where the subdivisions are areal spatial units, for instance, administrative grids), the graphic transcription of absolute quantitative data generally takes the form of a variation in the surface area of a simple point (proportional map). Cartograms (see Chapter 7) can also be used in certain restricted circumstances. There are however other possibilities.

Bertin's dots method consists in distributing dots evenly in each geometrical unit and varying their surface so as to directly represent a quantity. With this type of semiology, quantity is not transcribed by a proportional symbol positioned on a single point (centroid or particular place) but by a multitude of dots distributed evenly in each polygon. It is the sum of the surface area of the symbols that expressed the quantity in visual manner.

Dot density maps are maps where a scatter of dots known also as "countable dots" are randomly distributed within a unit. Each dot corresponds to a previously defined quantity, for instance, one dot for 1,000 inhabitants. These maps are constructed from absolute quantitative data across a grid that is as detailed as possible. Indeed, as the dots are distributed randomly in each grid unit, the smaller the grid is, the more precise will the dots be located.

These maps, often used in world atlases to represent population distributions, are an attempt to produce images that are "close to reality" and easily comprehensible by a user with no knowledge in statistics or geography. They are indeed easy to read, but fairly complicated to produce, the main challenge being localizing dots as precisely as possible (for instance, near coastlines or roads) to render the image relevant and comprehensible.

● **FOCUS: Did You Know?**

The first dot map was produced in 1830 by the Frenchman A.J. Frère de Montizon. This map, entitled "Philosophical Map" showing the population of France is the first map to represent absolute quantitative data.
⮑ https://frama.link/montizon

These two types of map, rather than being direct transcriptions of the property of proportionality, provide an idea of order and density. Although suited to absolute quantitative data, the eye will tend to apprehend the spatial structures and an implicit order, and not really the quantities represented. Thus, the areas of application are fairly limited.

**Figure 5.09**    Mapping absolute quantitative data.

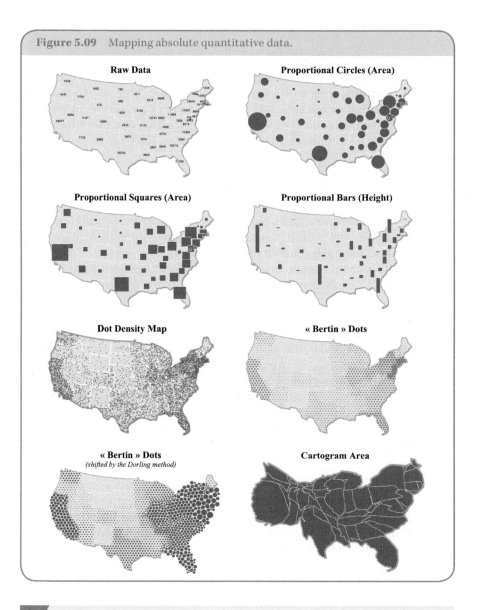

## Quiz

- What is a dot density map?
- Should values be discretized before using the visual variable "size"?
- How should we determine the anchorage point for a proportional symbol?

Chapter 6

# Solutions in Statistical Mapping

**Objectives**

- Knowing how to map temporal data
- Knowing how to design maps that can be compared
- Knowing how to map the results of bivariate and multivariate analyses.

Maps are remarkable tools for exploration and the prime tool of the geographer. A map makes it possible to spatialize, cross, and depict data and to reveal spatial configurations. It is an efficient tool for the appraisal and analysis of space. This chapter, rather than setting out to provide a theoretical framework, is intended to be read as a practical guide providing mapping solutions to the different situations commonly encountered when analyzing geographical data – mapping data at different moments in time, comparing two phenomena, and creating and mapping typologies. Readers should find ideas and useful recommendations.

## 6.1 MAPPING TIME

How can movements and trajectories in space or the evolution of a phenomenon over time be represented? The relevant modes of representation will depend on the scale on which the phenomenon is measured and on its representation (the map format).

**An individual trajectory** can be efficiently represented by a straightforward sinuous line on a two-dimensional map. The map will be successful if the line drawn faithfully represents the actual trajectory pursued. Significant place names can be marked on the map, so that the user can localize different elements. On these itinerary maps, temporal information (times of mobility and immobility) can be incorporated by adding a third dimension (space-time cube).

To represent **the evolution of a socio-economic phenomenon on an administrative grid** (for instance, population growth between two dates), the initial data needs to be converted. A growth rate will express the relative evolution of the phenomenon; a raw difference will express the absolute evolution. In these two instances, the graphic semiology used will need to suit the nature of the new data created in this manner. It should however be noted that this type of map, which is based on only two dates (start and end), is highly sensitive to the temporal options chosen. Any alteration, even if it is only small, of the starting and end dates will express the phenomenon in a very different manner. Often, the best way to apprehend spatio-temporal phenomena is to provide the user with several maps. These maps can be animated and succeed one another as in a film showing the evolution over time of the phenomenon.... More simply, the maps can be presented together, one beside the other, forming temporal sequence.

In the case of **a proportional symbols map** (for absolute quantitative data), you will need to ensure the comparability of the size of the symbols from one map to another. Two symbols of identical size (surface area, thickness, or height) should represent the same quantity. There should be a common legend for all the maps. In the legend, the smallest and largest symbols will correspond to the minimum and maximum values for all the maps involved.

If you develop two or more choropleth maps (relative quantitative data), the different data series considered as a whole should be discretized. In a spreadsheet-type software, a new statistical data series corresponding to the data for both or all the maps is thus be formed. Its analysis will enable the most relevant discretization method to be determined (see Chapter 2). Here again, the legend should be common to all the maps.

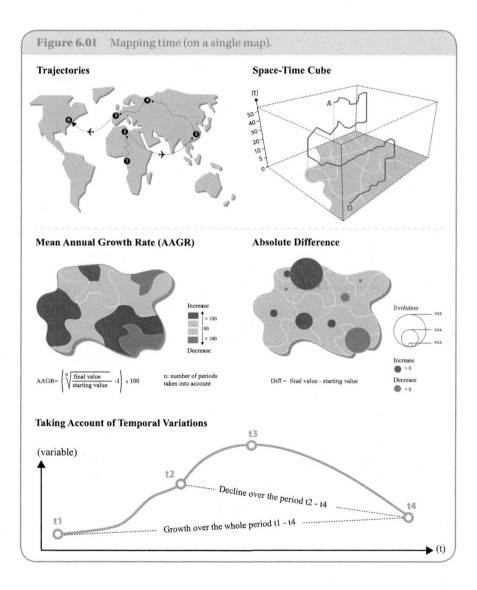

**Figure 6.01**    Mapping time (on a single map).

**Trajectories**

**Space-Time Cube**

**Mean Annual Growth Rate (AAGR)**

Increase

> 100

100

< 100

Decrease

$$\text{AAGR} = \left( \sqrt[n]{\frac{\text{final value}}{\text{starting value}}} - 1 \right) \times 100$$

n: number of periods
taken into account

**Absolute Difference**

Evolution

xxx

xxx

xxx

Increase

> 0

Decrease

< 0

Diff = final value - starting value

**Taking Account of Temporal Variations**

(variable)

t3

t2

Decline over the period t2 - t4

t1

Growth over the whole period t1 - t4

t4

(t)

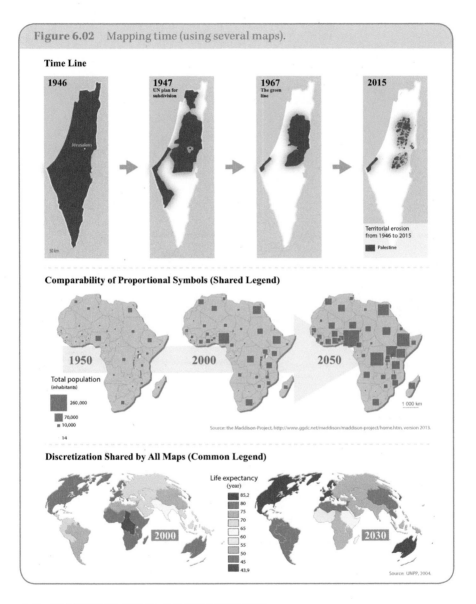

**Figure 6.02**    Mapping time (using several maps).

**Time Line**

1946

1947
UN plan for
subdivision

1967
The green
line

2015

Territorial erosion
from 1946 to 2015

Palestine

50 km

Jérusalem

**Comparability of Proportional Symbols (Shared Legend)**

1950

2000

2050

Total population
(inhabitants)

260,000

70,000

10,000

14

1 000 km

Source: the Maddison-Project, http://www.ggdc.net/maddison/maddison-project/home.htm, version 2013.

**Discretization Shared by All Maps (Common Legend)**

Life expectancy
(year)

85,2
80
75
70
65
60
55
50
45
43,9

2000

2030

Source: UNPP, 2004.

## 6.2 MAPPING FOR COMPARISON

How can we compare two geographical phenomena? There are various possibilities, depending on the data used.

The comparison of two sets of absolute quantitative data raises the issues of the order of magnitude and the measurement unit. How can we confront two maps of the world representing on the one hand GDP expressed in dollars and on the other the population expressed in numbers of inhabitants? One solution consists in expressing these two variables in proportions (percentages) of the world total. Once converted, the values of the two variables are expressed according to the same order of magnitude. In both cases, the total of the values is 100. Thus, the proportional symbols for the two variables become comparable. If the number of territorial units is not too large, the two types of information can cohabit on the same map (as is the case for "mushroom" maps).

If the aim is to set side by side two choropleth maps developed from different sets of relative quantitative data, the question of subdivision into classes is crucial. If the two distributions are symmetrical, a standardized discretization procedure (based on mean and standard deviation) can be used. If at least one of the distributions is not symmetrical, discretization based solely on the statistical number of individuals should be used: equal numbers.

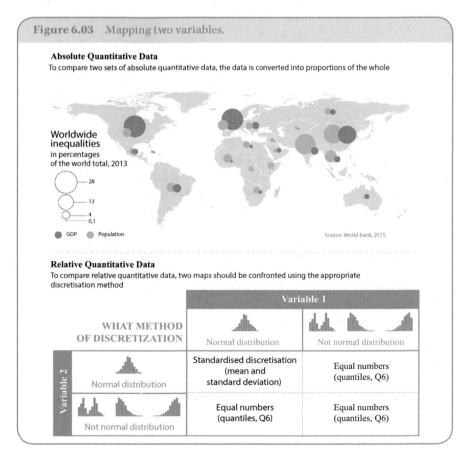

**Figure 6.03**   Mapping two variables.

**Absolute Quantitative Data**
To compare two sets of absolute quantitative data, the data is converted into proportions of the whole

Worldwide
inequalities
in percentages
of the world total, 2013

28
13
4
0,1

● GDP    ● Population                                    Source: World bank, 2015.

**Relative Quantitative Data**
To compare relative quantitative data, two maps should be confronted using the appropriate discretisation method

| WHAT METHOD OF DISCRETIZATION | Variable 1 | |
|---|---|---|
| | Normal distribution | Not normal distribution |
| Variable 2 — Normal distribution | Standardised discretisation (mean and standard deviation) | Equal numbers (quantiles, Q6) |
| Variable 2 — Not normal distribution | Equal numbers (quantiles, Q6) | Equal numbers (quantiles, Q6) |

## 6.3 MAPPING A RELATIONSHIP

One process that is widely used in geography consists in studying (and measuring) the relationship between two phenomena. This relationship can be apprehended in cartographic form by superimposing two or more phenomena. The relationship is then observed directly on the map by the superimposition of spatial configurations.

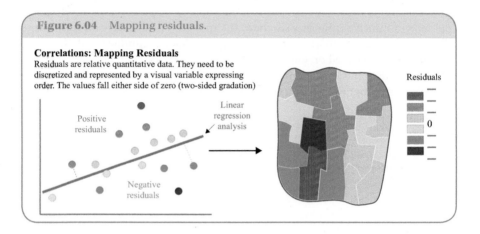

**Figure 6.04**    Mapping residuals.

**Correlations: Mapping Residuals**
Residuals are relative quantitative data. They need to be discretized and represented by a visual variable expressing order. The values fall either side of zero (two-sided gradation)

Positive residuals

Linear regression analysis

Negative residuals

Residuals

0

The relationship can also be envisaged in statistical form. This type of analysis, known as bivariate (two variables) or linear regression, enables a model to be derived determining the existence and the degree of correlation between two geographical phenomena. Once the model is established, mapping residues, which consists in assessing deviation from this model, makes it possible to reflect spatial units where the values diverge to a greater or lesser extent from what can be expected from the model. The residues are assimilated to relative quantitative data which need to be mapped as such (see Chapter 4). As some residues will be positive (above those in the model) and others negative (below those in the model), a two-sided color gradation will be used to efficiently reflect this contrast.

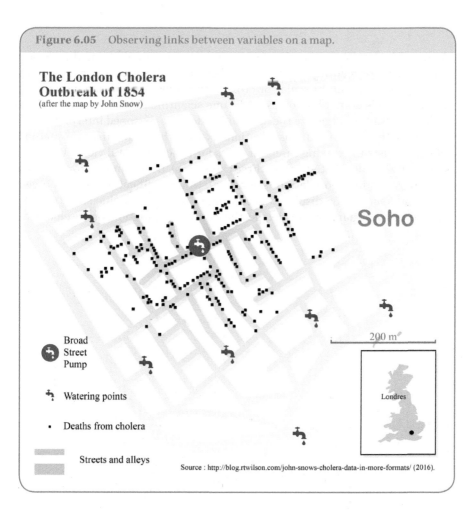

Figure 6.05    Observing links between variables on a map.

**The London Cholera Outbreak of 1854**
(after the map by John Snow)

Soho

Broad Street Pump

Watering points

· Deaths from cholera

Streets and alleys

200 m²

Londres

Source : http://blog.rtwilson.com/john-snows-cholera-data-in-more-formats/ (2016).

● **FOCUS: Correlation Does Not Mean Links!**

The correlation coefficient is a statistical measure enabling the study of a possible relationship between two quantitative variables. But it is not because two indicators are correlated that they are linked one to the other. A strong correlation coefficient does not establish a cause–effect relationship.

It is thus possible to establish links that are absurd: the number of individuals who died after falling into a swimming pool exhibits a 66% correlation with the appearance of Nicolas Cage in films; the consumption of cheese has a more than 94% correlation with the number of people who died entangled in their sheets. Watch out for spurious interpretations! ⟳https://frama.link/spurious

## 6.4 MAPPING TYPOLOGIES

Geographical information can be confusing. To render it intelligible and provide keys for reading, typologies are formed. In this area, there are numerous possibilities – multi-scalar or crossed typologies, composite indices, etc. These methods do not all possess the same scientific robustness. But they do make it possible to organize the data and to derive a reasoned interpretation.

One of the most used classifications is the triangular diagram method, which consists in crossing three variables for which the total yields 100% (the same variable distributed across three categories). If it is applied, for instance, to age groups in a population (see below), this classification enables a geographical space to be displayed according to its demographic structure: the over-representation of young people, the working population, or the elderly.

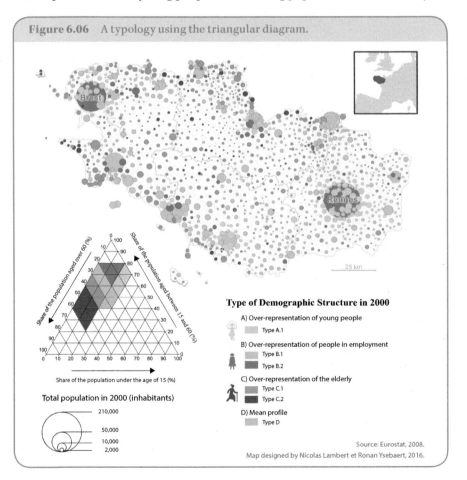

**Figure 6.06**    A typology using the triangular diagram.

Type of Demographic Structure in 2000

A) Over-representation of young people
  Type A.1

B) Over-representation of people in employment
  Type B.1
  Type B.2

C) Over-representation of the elderly
  Type C.1
  Type C.2

D) Mean profile
  Type D

Total population in 2000 (inhabitants)
210,000
50,000
10,000
2,000

Share of the population over 60 (%)
Share of the population aged between 15 and 60 (%)
Share of the population under the age of 15 (%)

Source: Eurostat, 2008.
Map designed by Nicolas Lambert et Ronan Ysebaert, 2016.

Another approach is apprehending the organization of geographical space according to different hierarchical levels at the same time: this is a multi-scalar typology. This method is based on the comparison of the value of each territorial unit according to three different contexts: a global context (the whole of the space under study), an intermediate context (the country or territory of which it is part), and a local context (the neighboring regions, contiguity, distance, etc.).

It is thus possible to position each territorial unit according to its position in relation to these three references. Careful choice of colors will then show up the particular situation of each territorial unit on the map. When applied, for instance, to per capita GDP in Europe, this method is particularly useful. It enables the user to see whether a region is more or less wealthy than the European mean, more or less wealthy than other regions in its member state, and more or less wealthy than its neighboring regions. A region can at once be considered wealthy (in relation to the situation in other regions in the same country) and poor (in relation to the European regions as a whole). Everything is relative....

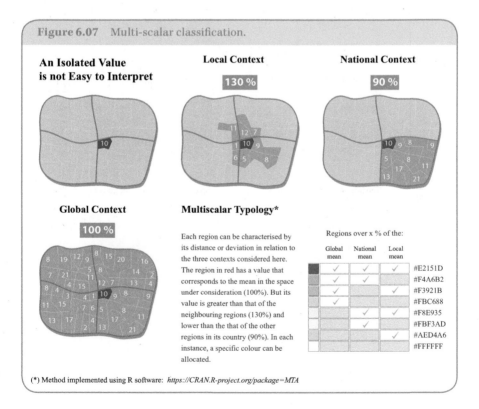

**Figure 6.07**    Multi-scalar classification.

**An Isolated Value is not Easy to Interpret**

**Local Context**    130 %

**National Context**    90 %

**Global Context**    100 %

**Multiscalar Typology\***

Each region can be characterised by its distance or deviation in relation to the three contexts considered here. The region in red has a value that corresponds to the mean in the space under consideration (100%). But its value is greater than that of the neighbouring regions (130%) and lower than the that of the other regions in its country (90%). In each instance, a specific colour can be allocated.

Regions over x % of the:

| Global mean | National mean | Local mean | |
|---|---|---|---|
| ✓ | ✓ | ✓ | #E2151D |
| ✓ | ✓ | | #F4A6B2 |
| ✓ | | ✓ | #F3921B |
| ✓ | | | #FBC688 |
| | ✓ | ✓ | #F8E935 |
| | ✓ | | #FBF3AD |
| | | ✓ | #AED4A6 |
| | | | #FFFFFF |

(\*) Method implemented using R software: *https://CRAN.R-project.org/package=MTA*

**Figure 6.08**   Unemployment in the USA according to three spatial contexts, 2008.

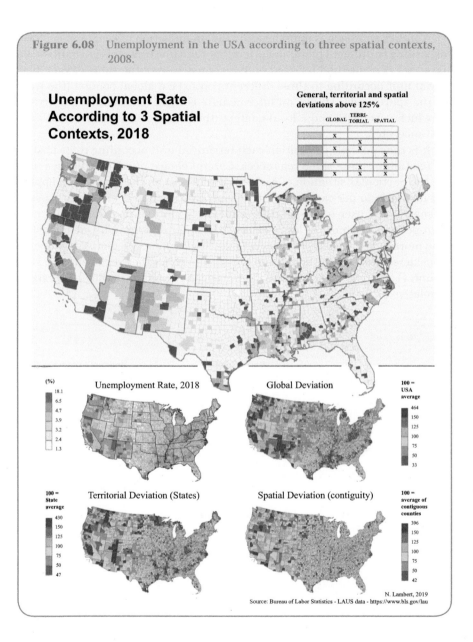

## 6.5 MULTIVARIATE MAPPING

Multivariate analyses are among the methods and tools that are widely taught and used in geography. They are based on statistical operations of varying complexity, not within the scope of this handbook. These analyses aim to sum up a large set of variables relating to geographical space. Most used in geography are the hierarchical clustering also known as hierarchical cluster analysis (divisive or agglomerative), principal component analysis (PCA), or correspondence factor analysis (CFA). Each method has its specific features, advantages, and drawbacks. While these methods are widely detailed in data analysis manuals (see bibliography), the question of their use in maps is rarely approached. We here provide a few rapid practical tips.

### 6.5.1 Agglomerative Hierarchical Clustering (AHC) Analysis

Agglomerative hierarchical clustering (AHC) analyses are based on a measure of dissimilarity (distance). They aim to form clusters of merged geographical units that resemble one another more than they resemble the units in the other classes. It's a bottom-up approach. When an AHC is run using statistical software, a cluster dendrogram is derived to reflect the proximity between the different classes formed by calculations of the distance between geographical units and a procedure of aggregation of these units into classes.

Mapping the results of an AHC should therefore lead to a map that primarily expresses differences between classes (color visual variable). The proximity or distance between classes is then transcribed by proximity between the colors used. Two colors that are close will be used to describe two classes that are close, and contrasting colors will be used to represent greater dissimilarities. This choice of colors, which is a step that is not easy to automate, is often not an easy task.

## Figure 6.09    Multivariate analysis and hierarchical cluster analysis.

### Hierarchical Cluster Analysis (HCA)

Mapping the results of this type of classification is performed on the basis of the table showing the profiles of the classes. Care should be taken designing the legend, and the page layout is also important.

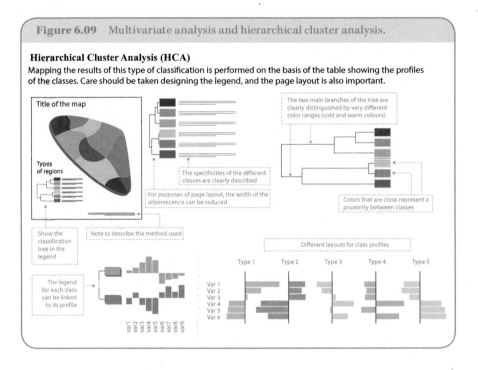

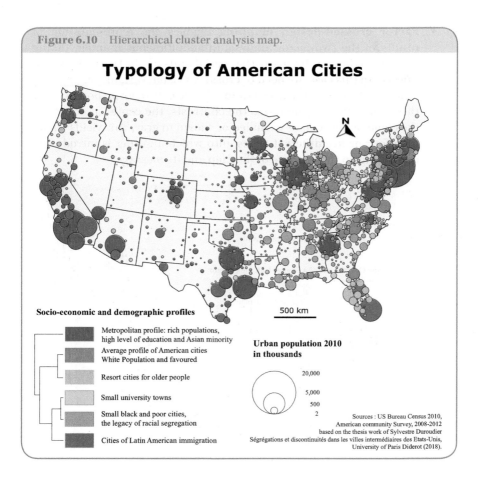

**Figure 6.10**    Hierarchical cluster analysis map.

# Typology of American Cities

**Socio-economic and demographic profiles**

Metropolitan profile: rich populations,
high level of education and Asian minority

Average profile of American cities
White Population and favoured

Resort cities for older people

Small university towns

Small black and poor cities,
the legacy of racial segregation

Cities of Latin American immigration

500 km

**Urban population 2010
in thousands**

20,000

5,000

500

2

Sources : US Bureau Census 2010,
American community Survey, 2008-2012
based on the thesis work of Sylvestre Duroudier
Ségrégations et discontinuités dans les villes intermédiaires des Etats-Unis,
University of Paris Diderot (2018).

## 6.5.2 Factor Analyses

Factor analyses are a family of methods enabling a dataset to be summed up using a restricted number of variables. The results of these analyses take the form of new variables (the factors) which are processed as relative quantitative variables, using a model that relates the new variables two by two, making it possible to use them as a typology.

Thus factor analyses enable the production of two types of map:

- **A factor map** (the factors being new statistical variables), where each factor is mapped independently according to the coordinates of the spatial units. The use of a two-sided gradation of colors enables the representation of positive and negative values, of dimensions contrasting one with the other.

- **Factor crossed-design mapping** enables the representation of results of the two-factor analysis in simultaneous manner, taking account of the coordinates of the spatial units for the two factors considered. In this case, we need to cross two visual variables expressing order (e.g., color and texture) to reflect the contrasts on both axes at the same time. To emphasize the contrasts, the center of gravity of the cloud of dots can be excluded or shown in gray. This yields a map with four or five different types, into which an order is introduced, depending on the purpose of the map and its message.

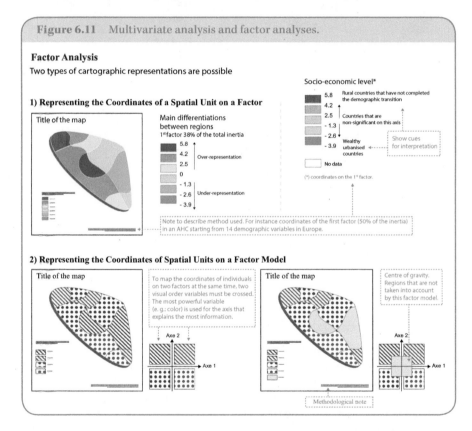

**Figure 6.11**    Multivariate analysis and factor analyses.

Mapping the results of multivariate analyses is not without its problems when it comes to designing the legends. Since the analyses are complex for non-specialists, care is required in designing the legends. A legend should enable the observation of phenomena that are evidenced in the statistical analysis in simple and direct manner. Here even more than elsewhere, the profiles of the future map users need to be specifically targeted.

**Figure 6.12**    Factor analysis map.

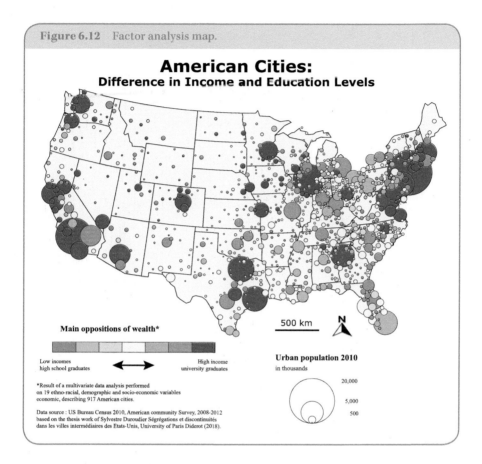

# American Cities:
## Difference in Income and Education Levels

**Main oppositions of wealth***

Low incomes
high school graduates

High income
university graduates

*Result of a multivariate data analysis performed
on 19 ethno-racial, demographic and socio-economic variables
economic, describing 917 American cities.

Data source : US Bureau Census 2010, American community Survey, 2008-2012
based on the thesis work of Sylvestre Duroudier Ségrégations et discontinuités
dans les villes intermédiaires des Etats-Unis, University of Paris Diderot (2018).

500 km    N

**Urban population 2010**
in thousands

20,000

5,000

500

## Quiz

- What discretization methods should be chosen to compare two maps?
- What is a multi-scalar typology?
- What are the two ways to map a factor analysis?

# Conclusion: Combining Visual Variables

Chapters 3–5 have sought to explain the logics of the graphic language used in cartography. Visual variables are the main components of this language, and correct use is fundamental for the effectiveness and visual perception of the chosen message. Designing a thematic map is governed by the use of one or several visual variables, which are always the expression of one or several pieces of statistical information (the data).

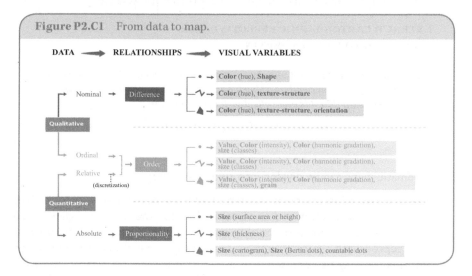

**Figure P2.C1** From data to map.

Chapter 6 focused on particular statistical procedures that any geographer will encounter frequently. These procedures, of varying complexity, can yield straightforward results that are fairly easy to represent. Certain other results can lead to the need to combine two or several visual variables in one and the same cartographic representation. Combinations of this sort enable the simultaneous transcription of several statistical data, which, once they are on the map, enables the different phenomena to be linked to each other visually. To do this, the use of different graphic figurations (points, lines, polygons) enables the superimposition of objects on one and the same map (for instance, roads and the administrative grid). For zone objects, the ground can allow for superimposition by way of a property of transparency. Today, the transparency option available on software also enables the superimposition of opaque grounds. Care is however required with respect to the final effect of the colors mixed in this way. Finally, some combinations of visual variables do not raise superimposition issues since they act of different parameters (for instance, color and shape). Certain associations are in fact very effective (for instance, size+color).

While it is common to use combinations and superimpositions of visual variables to transcribe different types of information, these variables can also be used to "say" the same thing twice in two different ways. Graphic redundancy, consisting in using at least two visual variables to represent the same statistical information several times, is a means of making the map more readable. In the same way as for oral or written communication, it can be useful to explain the same thing several times in different ways so as to ensure the idea is understood. A map can also make use of this rhetorical device.

**EXTRACT: CARTOGRAPHIC EFFECTIVENESS ACCORDING TO JACQUES BERTIN**

*If to obtain a satisfactory and complete response to a given question, all things being equal, one construction requires a shorter observation time than another construction, it can be said that it is more effective for this question.*

*(1967 Sémiologie Graphique, les diagrammes, les réseaux, les cartes)*

● **FOCUS: Maps to Be Read or to Be Looked At?**

Thematic maps can be divided into two categories:

**Maps to be read** are maps the aim of which is not to immediately deliver a message that is clear and comprehensible for the user at a glance. On maps intended to be read, the different elements of geographical information represented are often juxtaposed, and the spatial organization of the information may be confused. It is a map that is read like a table, and each piece of information will be sought individually. Maps to be read enable the display of precise information on a place, the comparison of two places, but they do not readily provide a view of the global spatial structure. These are maps that are read place by place.

**Maps to be looked at** for their part are constructed on the idea that the message delivered by the map should be understood by the user in less than ten seconds, considered by Bertin to be the minimum viewing time required. To construct this type of map, careful design of the hierarchization of the information to be delivered is required, with a skillful use of graphic semiology and verbal components with maximum impact. This type of map is didactic; it is intended to explain a geographical phenomenon to the viewer.

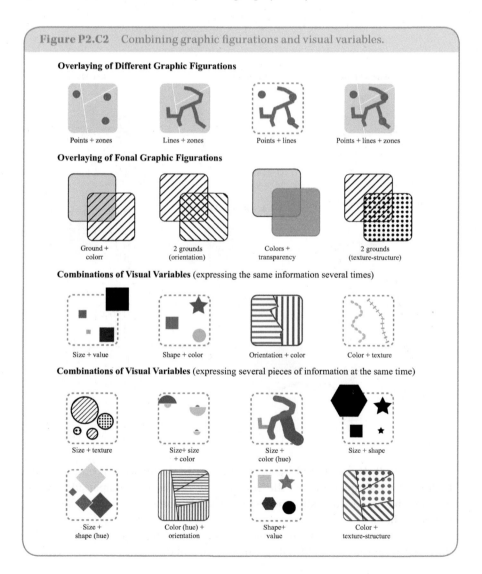

**Figure P2.C2**   Combining graphic figurations and visual variables.

Care is however required to preserve readability. The superimposition of information has an impact on the clarity of the message delivered by the map. A map superimposing a large amount of information less easily delivers a message that is readily clear and comprehensible. We are talking here more about "maps to be read" where the aim is not to provide an immediate visual message. Fortunately, when the information is complex, there are solutions other than superimposition.

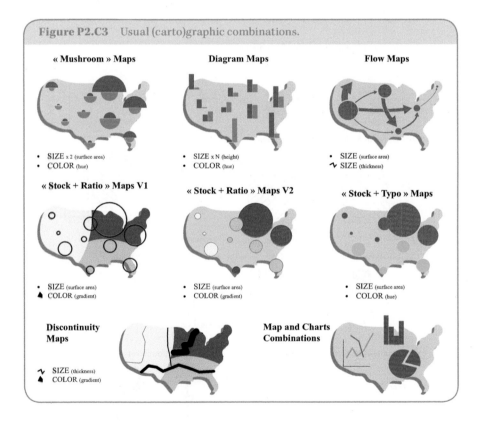

**Figure P2.C3**    Usual (carto)graphic combinations.

First of all, there should be no hesitation in duplicating a complex map to form two simpler maps. The positioning of the two maps side by side enables the reader to easily understand the phenomena and the relationship between them.

It can also be effective in some cases to combine a map with another, non-cartographic element (histogram, curve, and diagram).

Finally, the simplification of complex data does not solely involve the use of graphic elements. It can also be performed upstream, by statistical processes and classifications of one or several variables (see Chapter 6). To do this, there are numerous methods enabling the reduction of the information to be represented (composite indices, bivariate or multivariate analyses, etc.).

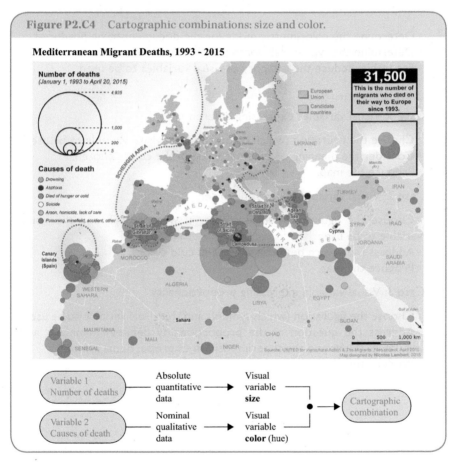

**Figure P2.C4**    Cartographic combinations: size and color.

*The map of migrants who died on the borders of Europe was designed in 2003 by Olivier Clochard and then revised by Philippe Rekacewicz for a first publication in* Le Monde Diplomatique. *Since then, there have been numerous updates. These successive maps have enabled the displacements of migrant flows to be evidenced as well as the increase in the numbers of deaths over the years. On this map, two variables are used: the number of dead (absolute quantitative data expressed by the visual variable "size") and cause of death (nominal qualitative data) expressed by the visual variable "color". While this map has become a "classic", other modes of representation have recently been implemented from this data.*

⮑ *https://frama.link/theborderkills & https://frama.link/theborderkills2*

## THE MAP GAME PART 2: DESIGNING A MAP

The map to be produced is in color and intended for display on a screen with a format 800 pixels wide at most

### I. Initial choices

• Determine the colorimetric mode for the document to be produced.
• Determine the type of relationship with the data chosen.
• From this, deduce the category of visual variables to be used.

### 2. Implementation

Case I: Visual variables expressing differentiation

• Choose the figuration (which colors) and the visual variable to be used.
• Consider the meaning or symbolic value of the colors chosen.
• Produce the map.

Case 2: Visual variables expressing order.

• Choose the figuration (which colors, in what order?) and the visual variable to be used.
• Address the choice between warm and cold colors.
• Produce the map.

Case 3: Visual variable expressing proportionality.

• Choose the figuration (which symbol) and the visual variable to be used.
• Address the question of the proportionality between minimum and maximum values and the superimposition of proportional symbols.
• Produce the map.

# Part 3

# Beyond the Visual Variables

*Mind you, I'm not sure that reality is enough. Even if reality actually exists, and can be reproduced or reflected, reality, as you say, has to be transformed, interpreted. Without a way of looking at things, without a viewpoint, at best it's eternally boring, at worst it's unsettling. And the process of transformation, whatever the material you start off from, is always a fiction.*

Delphine de Vigan, *D'après une histoire vraie (Based on a True Story)*, Jean-Claude Lattès, 2015

## INTRODUCTION

Mapping means showing facts and telling a narrative about space. It can be basically a purely technical action, yet it is above all an act of creation. To produce a map is to construct a world made up of knowledge and science, but also of dreams and poetry. Kandinsky, in 1911, said that "to create a work of art is to create a world". A map leads you at once to dream and to think. By the choice of evocative colors, poetic place names, winding rather than straight lines, and imaginative titles, a map takes you on a journey. It is a motionless journey, where the spatial configuration comes to life by way of the mapmaker's imagination, offering the reader a map that can be perused in three dimensions.

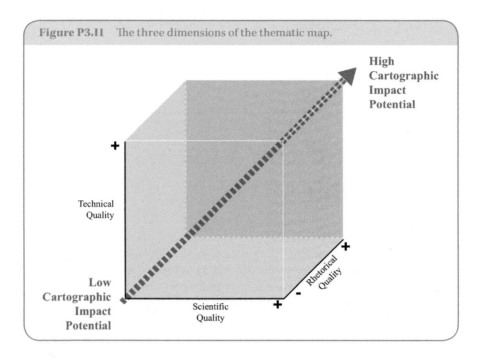

**Figure P3.I1**    The three dimensions of the thematic map.

A map is a **scientific** object. From the statistical processing through to its graphic figuration, there are precise rules that need to be strictly complied with. Statistical rules first of all – the need to reduce information entails particular rules enabling the underlying organization of the data to be accurately represented. Next, there are the semiological rules: cartography has a grammar – the visual variables – which need to be handled correctly. Finally, creating a map also requires knowledge, documentation, and geographical concepts that will need to be called upon to evidence spatial organizations that have meaning.

The map is also a **technical** object. It results from a succession of phases or procedures that are more or less reproducible and enable the conversion of raw data into a geographical image. Most of the time, this conversion process is performed using computer tools. But it can also be done manually: pencil, paintbrush, freehand sketches, etc. In this second approach – which is artistic – while there is a loss in terms of reproducibility, there is likely to be a gain in terms of originality and aesthetics.

Finally, the map is a **rhetorical** object. Every map conveys a narrative which is expressed via the choice of colors, the outlines (or absence thereof) of borders, the lexical elements to include, the titles, the hierarchy of the information, the exaggerations, and the omissions. In cartography as in music, there is considerable scope for interpretation, and as in music, there are also different styles of mapping. A map can have a head-on impact like heavy metal; it can be sophisticated like classical music, unexpected and iconoclastic like free jazz, or rebellious like angry rap. In fact, a map can also be appraised by the effectiveness of its rhetoric, that is to say its ability to capture the viewer's attention to "tell him a story".

After presenting a few original cartographic representations (Chapter 7) and after detailing different ways to "dress" the map (Chapter 8), this third part will go on to show how, beyond the scientific and technical approaches, any map can take on several faces and different narratives (Chapter 9).

● FOCUS: Can a Map Be Made to Lie?

In 1991, the Mark Monmonier published a book entitled *How to Lie with Maps*. For this author, not only is lying easy with maps, it is also essential. He adds that "most decent maps are a collection of little lies" and those white lies are essential elements in the mapmaker's language.

So do maps really lie? In all events, if they do, they can lie honestly. Maps should be conceived like mirrors than enlarge rather than distort. Their purpose is to cast light on a facet of reality, but not to deform reality to provide a misleading image. The task is nevertheless a delicate one. If it enlarges some elements and reduces others, a map provides only a partial image of the world. Thus, the aim should be to use distortions that convey meaning, based on scientific reasoning and established facts. Maps lie only if the mapmaker intentionally deforms reality to show something that he or she knows to be false. Otherwise, maps are partial, subjective contributions to the understanding of a complex, plural reality, sometimes in contradiction with other proposals. The aim of mapping is not to deform reality but to make it more intelligible, which inevitably entails simplification.

# Chapter 7

# Other Cartographic Presentations

**Objectives**

- Knowing how to construct a grid-referenced map
- Knowing how to produce a smoothed map
- Knowing how to map discontinuities
- Knowing how to construct a flow map
- Knowing how to construct a cartogram.

Beyond the mere application of a visual variable to a punctual, linear, or areal spatial unit, which belongs to the area of graphic semiology, mapping can also be seen in terms of its mapping or designing processes. These processes, which run from the raw data to its graphic expression, can be described in a logical series of transformations that are traceable and reproducible. These operations, which can be automated, sum up the way in which a thematic map is designed. These procedures, which can be easy or relatively difficult to implement on cartographic software, can amount to a straightforward processing of the data and the use of a visual variable (choropleth maps; typologies), or require complex processes of conversion (grid referencing, smoothing). Certain types of conversion can also lead to changes in the form that the map will take on, altering contours, boundaries, and positions (this is the case with cartograms).

● **FOCUS: Reproducibility**

In the field of research, reproducibility means that each scientific publication is accompanied by datasets and source codes enabling fellow researchers to reproduce the results of an analysis, so as to validate or contest them. In 1889, Anatole France, in his novel *Balthasar*, wrote: "Science is infallible; but scientists are always making mistakes". Cartographers can get it wrong too. Like researchers, they therefore have everything to gain from publishing their data and specifying the software used along with the image produced.

Far from covering all these instances, we choose, in this chapter, to detail five types of map, five modes of representation frequently used for thematic maps. These five cartographic "faces", to recall Roger Brunet's "Visages de la carte" (1990), each have a specific usefulness and show geographical space according to different, complementary facets (continuum, interruption, interaction, etc.).

● **FOCUS: Multi-Representation in Mapping**

Geographical reality is complex and often contradictory. It can never be mapped in exhaustive manner. In order to reflect its complexity, a single cartographic representation is sometimes not enough. Creating and confronting several cartographic productions of one and the same phenomenon broadens the debate and enhances geographical knowledge.

## 7.1 REGULAR GRID MAPS

In the field of socio-economic thematic mapping, cartography is most often based on a particular grid reference system, the administrative divisions. It can seem perfectly reasonable to develop maps according to this system of subdivision on several counts. First of all, it is simpler – data and basemaps derived from this system of subdivision are directly available, and in addition, the territorial units identified in this way are familiar and recognizable, and they have names and real existence.

Yet while it appears logical to use this type of grid, the choice is not without meaning. Administrative grids are heterogeneous, and they alter over time. They act as "territorial filters" and have considerable impact on cartographic representation (see Chapter 2).

A grid-referenced map consists in subdividing geographical space into a grid made up of regular squares (or other simple geometric shapes), on a given projection. It thus escapes from the arbitrary and irregular nature of administrative subdivisions. It evidences major trends in spatial distributions derived from data by way of a subdivision into equal, localized squares. The data is distributed across this regular grid placed over the map in proportion to the surface area represented.

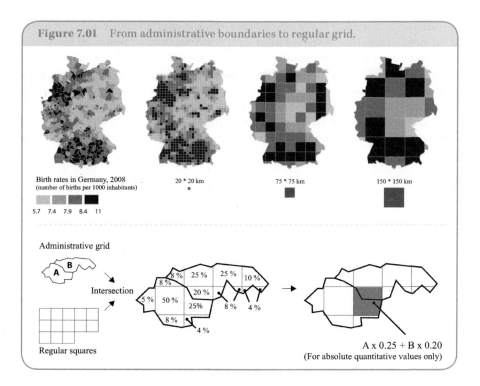

**Figure 7.01** From administrative boundaries to regular grid.

Birth rates in Germany, 2008
(number of births per 1000 inhabitants)

5.7  7.4  7.9  8.4  11

20 * 20 km

75 * 75 km

150 * 150 km

Administrative grid

B
A

Intersection

Regular squares

8 %  25 %   25 %  10 %
8 %
5 %  50 %   20 %
25%   8 %  4 %
8 %
4 %

A x 0.25 + B x 0.20
(For absolute quantitative values only)

● **FOCUS: 1789 – Square French** *départements*

Before the French revolution in 1789, the kingdom was divided into a multitude of provinces and other subdivisions under the *Ancien Régime*. In an
effort to reshape the territory, Jacques Guillaume Thouret (1746–1794)
delivered an address to the Assembly on September 29, 1789, in which he
proposed a new subdivision of the territory into departments in the form
of squares. According to this checkerboard subdivision, each department
was to correspond to a 72 km square, itself subdivided into smaller squares
forming districts and cantons.

Since the Assembly expressed worries about this type of subdivision, the pro-
posal was abandoned. On October 3, a map of the French *departments* was
drafted, on which the departments were not so geometrical, and for a large part
reverted to the former provincial boundaries. It is not easy to erase the past!
➲https://frama.link/squares

This technique has three advantages. First, it enables the creation of a neutral subdivision that does not alter over time. Second, a change in the size of the squares enables the production of images with a varying degree of generalization depending on what one is seeking to show. Finally, squared grids can also offer a solution for thematic harmonization, enabling the combination of data sources that are not normally compatible. With this method, for instance, it is possible to cross data relating to watersheds with data relating to the administrative regions, two systems of subdivision that are different in nature.

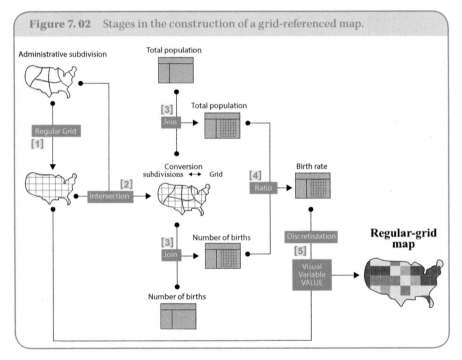

**Figure 7. 02**    Stages in the construction of a grid-referenced map.

*The production of a regular grid maps takes place in several stages which can be automated. A regular grid in the initial projection system is created to provide the base for the representation [1]. An intersection operation enables a conversion phase from an administrative subdivision to a regular grid [2]. A joining procedure enables data to be allocated to each square in the grid in proportion to the intersected surface area [3]. Once this is done, the map is constructed quite simply: creation of a ratio [4], discretization, application of a visual variable of the "value" type [5], page layout, etc.*

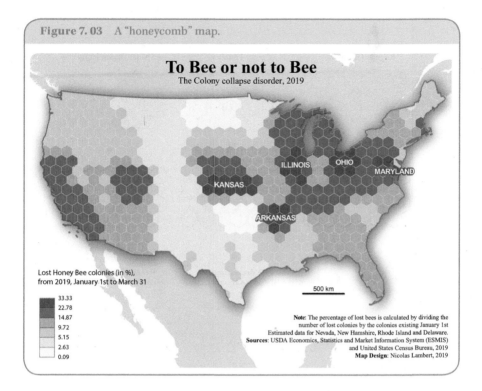

**Figure 7. 03**    A "honeycomb" map.

# To Bee or not to Bee
The Colony collapse disorder, 2019

ILLINOIS    OHIO    MARYLAND

KANSAS

ARKANSAS

Lost Honey Bee colonies (in %),
from 2019, January 1st to March 31

| | |
|---|---|
| | 33.33 |
| | 22.78 |
| | 14.87 |
| | 9.72 |
| | 5.15 |
| | 2.63 |
| | 0.09 |

500 km

**Note**: The percentage of lost bees is calculated by dividing the
number of lost colonies by the colonies existing January 1st
Estimated data for Nevada, New Hamshire, Rhode Island and Delaware.
**Sources**: USDA Economics, Statistics and Market Information System (ESMIS)
and United States Census Bureau, 2019
**Map Design**: Nicolas Lambert, 2019

## 7.2 DISCONTINUITIES

*A continuous natural world, pottering on without breaks or discontinuities, without
jumps or sudden changes, what a dull and flabby world it would be! A world in which
each phenomenon would evolve cozily. (Tamed Nature. Ugh!)*

Denis Guej, *Le théorème du perroquet (The Parrot's Theorem)*, 2000

While it is possible to measure disparity in a geographical phenomenon using
statistical measures (variation coefficients, Gini index, etc.), these measures
tell us nothing about their configuration in space. Yet the spatial organization of geographical phenomena does have meaning. Space can be organized
in different manners: continuous, discontinuous, polycentric, etc. Situations
are not always comparable. Marked spatial discontinuities are often areas
of considerable tension, and it is interesting to show this on the map of the
phenomenon at hand (inequalities in wealth, ghettoization, climatic discontinuities, etc.).

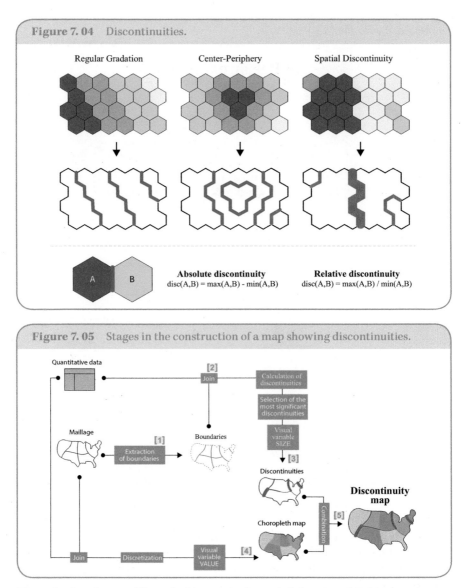

**Figure 7.04**    Discontinuities.

Regular Gradation            Center-Periphery            Spatial Discontinuity

**Absolute discontinuity**
disc(A,B) = max(A,B) - min(A,B)

**Relative discontinuity**
disc(A,B) = max(A,B) / min(A,B)

**Figure 7.05**    Stages in the construction of a map showing discontinuities.

*The production of a map of discontinuities starting from administrative subdivisions requires several steps. The first is to extract the boundaries between administrative units in the form of lines. Each line is then allocated two codes corresponding to the codes of the units on either side of the boundary [1]. A joining procedure then enables the data linked to the subdivisions to be allocated to the boundaries [2]. Before representing them using the visual variable "size" (thickness), the value of the discontinuity is calculated according to a defined method [3]. To reflect the meaning of the discontinuities, a choropleth map can be created [4] serving as the base for the representation.*

By tracing lines (boundaries) between administrative units that vary in thickness (visual variable "size"), discontinuity maps make it possible to put the focus on breaks in geographical space. When this is combined with the use of colored areas (choropleth map), the cartographic representation is particularly effective in producing an image showing major spatial breaks. This type of map draws the eye, not to the large homogeneous areas but rather to the boundaries evidencing marked disparities in the phenomenon under study on either side of a boundary. The value of disparities can be calculated in different ways. For greater clarity on the map, generally speaking only marked discontinuities are represented.

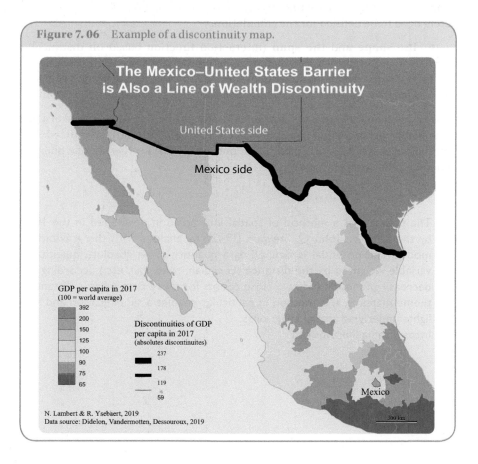

**Figure 7. 06**    Example of a discontinuity map.

## 7.3 SMOOTHING

Many socio-economic phenomena are spatially distributed according to a continuous logic, without major breaks at the boundaries, which then appear as artificial divisions. To stress continuous logics of this type, a mathematical procedure, the potential, makes it possible at any point in space to calculate the value of an absolute quantitative variable (stock) localized in its neighborhood. This semantic simplification removes the "noise" linked to the heterogeneity of the system of subdivision (producing statistical information from subdivisions of very varied size introduces interpretation biases, see Chapter 1) and highlights the general underlying structure of the territory. Several parameters need to be defined for the calculation:

- **The shape** and **the span** (interaction function based on distance) determine the way in which the neighborhood is taken into account
- **The slope** of the mathematical function (known as the friction of distance) has an effect on the degree to which distance is taken into account.

The variation of these parameters enables the production of an image of a geographical phenomenon that is more or less simplified and generalized. This method makes it possible to highlight both the local specificities of a phenomenon and the general trends.

● **FOCUS: The Potential**

The potential is a method of spatial interpolation developed in the 1940s by the physicist John Q. Stewart (1942), by analogy with the gravitational model. The potential is defined as a quantity of an absolute quantitative variable weighted by the distance (Euclidian, road, rail, etc.) according to a decreasing function (what is close counts for more than what is distant). In theme mapping, this method in particular enables a simplification and highlights underlying spatial structures.

https://frama.link/smooth

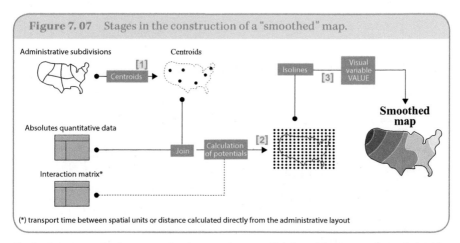

**Figure 7.07**    Stages in the construction of a "smoothed" map.

*Producing a smoothed map consists in several stages. The data (absolute quantitative) is distributed in space on point objects (for instance, the centroids of territorial units) [1]. Three elements are then required for the calculation of the potentials: a calculation grid, a distance (Euclidean, by road, by air, etc.), and an interaction function (and its parameters). The finer the grid, the longer the calculation, but the more precise the result [2]. A vectorization process, enabling the construction of isolines and surface areas for each set of values, is often used to finalize the cartographic representation [3].*

Just as a trend sums up a series of values collected over time in more comprehensible manner, a smoothed map makes it possible to focus only on the main phenomenon. When seen in 3D (see below), this type of map also makes it possible to reflect the intensity of the phenomenon represented by the height of the "peaks". Ultimately, smoothed maps produce synthetic images that are easy to understand and pleasant to look at.

**Figure 7. 08**    Example of smooth map.

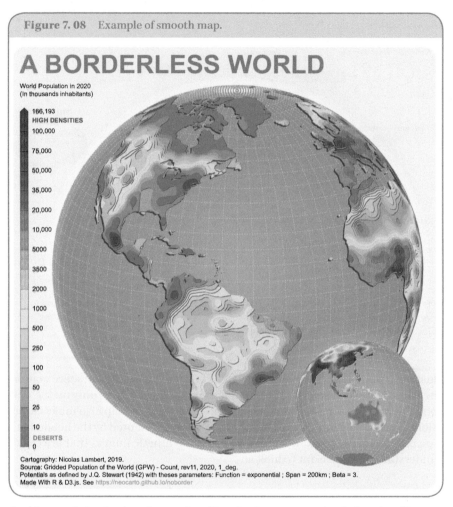

On this map, the borders have been replaced by other lines representing the location of humans on the earth's surface. This smoothed representation aims to show the continuums of the world and to get out of the habitual representations of an enclosed world within the grid cells of the states. This map shows also the world as a globe. A world without borders and without edges, to show, in the Elise Reclus manner, the idea of a united and solidary world (see https://neocarto. github.io/noborder).

## 7.4 CARTOGRAMS (OR ANAMORPHOSIS)

The term "anamorphosis", generally defined as a distortion of an image, is not specific to cartography. This procedure consists in producing images distorted in such a way that they recover their normal aspect when viewed from a certain distance and a certain angle, or when they are reflected in a concave mirror. It is used in numerous areas (painting, advertising, marketing, road signs, and even in psychoanalysis). The term "anamorphosis" is thus one that refers to a variety of practices, allied to optical illusions.

● **FOCUS: Anamorphosis in Painting**

In the arts, the use of anamorphosis was made famous by Holbein's painting "The Ambassadors", where a strange object can be seen. It proves to be a human skull when the painting is viewed not face on but from the side. In this painting dated 1533, incidentally, there is also Johan Schöner's terrestrial globe placed upside down on the left of the shelf.

In cartography, it is preferable to use the term "cartogram". A cartogram is the result of a distortion of the basemap in accordance with a quantity that is applied to the different geographical objects making it up. Cartograms can translate both quantities and links between places (for instance commuting times). In both cases, the outside contours on the map are altered.

● **FOCUS: The First Cartogram**

The first cartogram was produced in 1921 by General Electrics. It was a map of the United States where each state was enlarged or reduced in accordance with its electricity consumption. Before this map, other approaches had been attempted. The first was that of the French economist Emile Levasseur who in 1868 produced a map where each country was represented by a square proportional to its surface area.

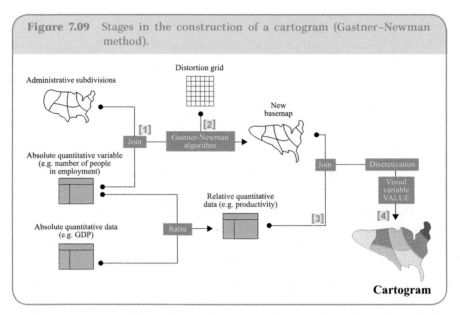

**Figure 7.09**    Stages in the construction of a cartogram (Gastner–Newman method).

*Producing a cartogram with the Gastner–Newman method starts from absolute quantitative data [1]. A calculation grid and a certain number of iterations are defined to determine the precision and the quality of the conversion [2]. The new, distorted basemap can then be used to produce a choropleth map (or other kind of representation) [3] and [4].*

There are today many methods of distortion, so that it is not possible to refer to the cartogram as a single object. Colette Calvin [2008] has indeed produced a detailed classification which she refers to as the "cartographic transformation of position". There is one method, nevertheless, that is widely dominant in contemporary mapmaking. Developed by Gastner and Newman in 2004, the method consists in distorting a system of subdivisions starting from absolute quantitative data, while at the same time preserving contiguities between these territorial units. This procedure consists in applying the visual variable "size" to surface area (zone) geographical objects.

**Figure 7.10    Different types of cartogram.**

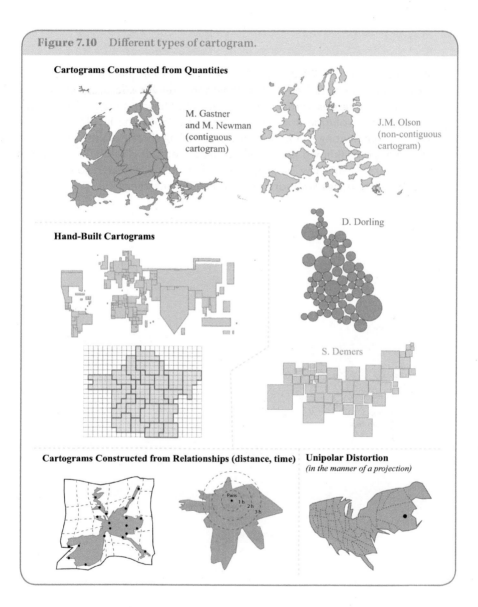

Cartograms Constructed from Quantities

M. Gastner and M. Newman (contiguous cartogram)

J.M. Olson (non-contiguous cartogram)

D. Dorling

Hand-Built Cartograms

S. Demers

Cartograms Constructed from Relationships (distance, time)

Unipolar Distortion
*(in the manner of a projection)*

Paris
1 h
2 h
3 h

● **FOCUS: Waldo Tobler, A Pioneer in**
**Computer-Assisted Cartography**

Waldo Tobler (1930–2018) was a Swiss-American. He was a precursor in the area of computer-assisted cartography, and after his doctorate in 1961, he developed numerous spatial analysis methods. He was a major figure in the quantitative revolution in geography from the 1950s and in 1973 published an article presenting the first algorithm enabling the automatic production of a cartogram (known as the rubber map method). Waldo Tobler is also famous for his "first law" in geography: "Everything is related to everything else, but near things are more related than distant things".

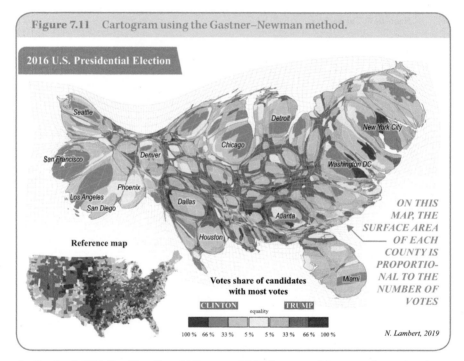

**Figure 7.11**    Cartogram using the Gastner–Newman method.

On October 3, 2019, President Donald Trump published a tweet with a map of counties almost all colored in red (resembling the reference map below) with the following legend: try to impeach this! Targeted by an impeachment procedure conducted by the democratic elected representatives, Trump then used the mapping argument to convince people of its legitimacy. But it's a cartographic illusion. By taking into account in the same way all the counties, whether they are totally empty or very densely populated, this map lies. By adding a size criterion based on the number of voters, the map radically changes its face. The red areas become less imposing, and large blue areas are immediately apparent. Let us remember that, in 2016, Donald Trump was elected with three million fewer votes than Hilary Clinton.

Cartograms, which are seen as innovations (despite the fact that they have been around for nearly a century), produce highly generalized images that are very efficient in representing quantities and gradients. They are maps that are effective in ordering geographical information in a global, structured image.

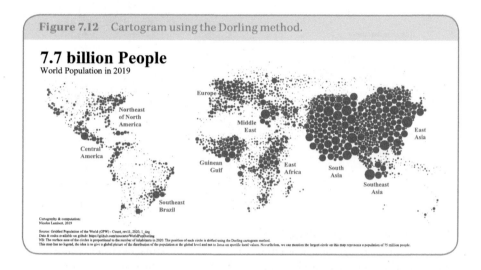

**Figure 7.12**    Cartogram using the Dorling method.

## 7.5 FLOW MAPS

Although they are not a major cartographic innovation, flow maps raise complicated, recurrent problems. They are not easy to produce on computer-assisted programs, but they are important, classic representations whenever the aim is to represent exchanges between geographical objects, both quantitative and qualitative.

● **FOCUS: The Spaghetti Effect**

In 2015, 197 countries were recognized by the United Nations Organization. A map showing all the flows between the different states would need to display 38,612 arrows representing the links of each country with the others across the world ($38,612 = 19 \times 197 - 197$). As these links would be too numerous, they cannot be represented simultaneously, because it would make the map illegible. To reduce this graphic complexity, a flow map therefore always requires a crucial stage of selection, aggregation, or hierarchization of the relevant information, this being the only way to avoid the map looking like a plate of spaghetti.

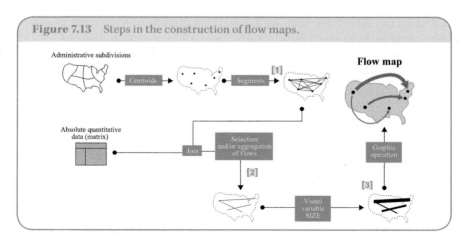

**Figure 7.13**    Steps in the construction of flow maps.

*After defining the links between pairs of places [1], the production of a flow map most of the time requires a selection or aggregation stage enabling the reduction of the number of lines to represent. This operation can be performed automatically, but it often requires a fairly detailed analysis of the data [2]. Finally, the application of the visual variable "size" completes the representation [3].*

The flows that are represented, most often by way of arrows from one spatial unit to another, provide information on relationships between places. The nature of the exchange can be represented by a color, the scale of the exchanges by the thickness of the shaft of the arrow, and the direction of the relationship by the direction of the arrow head. While it is fairly easy to automatically calculate lines of varying thickness between two geographical objects, the positioning of the flows one in relation to the other, the curve, and the management of superimpositions of the arrows are graphic elements that are difficult to automate efficiently. Thus, the cartographer will need to deploy his or her skills and know-how to arrange the arrows in logical manner, minimizing overlap, so as to produce an effective, intelligible image. A process of selection of the relevant information is generally required. For enhanced readability, it is also important to suit the cartographic projection used to the nature of the flows under study.

## ● FOCUS: The First Flow Maps

The first flow map is attributed to the Englishman Henry Drury Harness in 1837. In a study on the Irish railways, he produced a series of maps on the circulation of goods and passengers using proportional figurations. The procedure used for this map, which is closer to a network map (flow map on the use of the network), was returned to later by a Frenchman, Charles Joseph Minard, who in 1869 produced his famous diagram representing the evolution of numbers in Napoleon's army during the Russian campaign. Finally, in

1885, the German geographer Ernst Georg Ravenstein produced what was probably the first genuine flow map using arrows, on the mobility patterns of British people at county level.

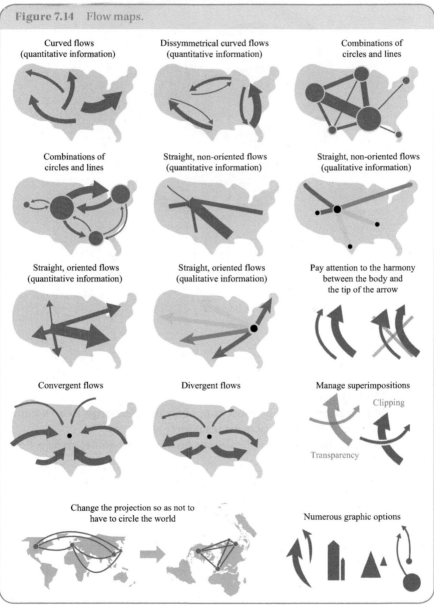

**Figure 7.14**    Flow maps.

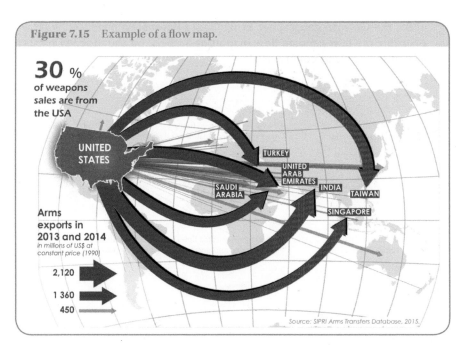

**Figure 7.15**    Example of a flow map.

30 %
of weapons
sales are from
the USA

UNITED
STATES

TURKEY

UNITED
ARAB
EMIRATES

SAUDI
ARABIA

INDIA

TAIWAN

SINGAPORE

Arms
exports in
2013 and 2014
in millions of US$ at
constant price (1990)

2,120

1 360

450

Source: SIPRI Arms Transfers Database, 2015.

*This flow map shows the countries that purchased weapons from the USA in 2013 and 2014. Over this period, six countries bought more than 1,360 million dollars' worth each. These large flows are represented on the map by thick, curved lines (so as to differentiate them) which are also opaque (to highlight them). The other more minor flows are marked as straight, transparent lines.*

## Quiz

- What is a potential?
- Which algorithm is the most commonly used to design a cartogram?
- What are discontinuity maps used for?
- What is the stage that should not be missed when designing a flow map?

# Chapter 8

# Staging

- Knowing how to "dress" a geographical image
- Knowing how to organize the elements making up the map
- Knowing how to bring the map to life by way of its "staging".

A map is only considered to be finished when the cartographic image has been set out on a page. This finalization phase consists in placing the different elements making up the map, taking account of legibility, and the size and format of the map. This is an essential step in the cartographic construction process. Its purpose is to organize the elements on the map and to add interpretation keys in such a way that the main message of the map is understood by the viewer in the first 10 seconds.

● **FOCUS: Maps on Screen and Maps on Paper**

**Paper maps:** The layout on the page can be either "landscape" or "portrait". While the landscape orientation is often used in atlases, the portrait orientation is the most frequent strategy. In all cases, the orientation of a document has an impact on the way in which the different graphic elements are positioned. As the map itself is the most important element, it should be as large as possible. Blank surface areas that are not used should be minimized. To be printed on paper, the map will need to be designed in CMYK mode (see Chapter 3) and exported with a resolution of over 300 dpi (digits per inch). When a map is to be included in a book, the author will also need to consider the positioning of the map in relation to the text around it, and how to articulate map and text, or map and figures or tables. If two graphic elements are to be read at the same time (for instance, map and legend), the map designer will need to ensure that this is indeed feasible. Page numbers also need to be managed.

**Maps on screen:** the format of maps intended for display on a screen is not expressed in centimeters or inches, but in numbers of pixels. The size available is thus determined by the website or the computer application. To be correctly displayed on a screen, a map needs to be designed in RGB mode (see Chapter 3) and exported at a resolution of 72 dpi (which corresponds to the resolution of computer and tablet screens). If the map is posted on a website, it may be useful to integrate all the elements making up the map within the image (title, legend, copyright, etc.) so that it cannot be trimmed or diffused only in part.

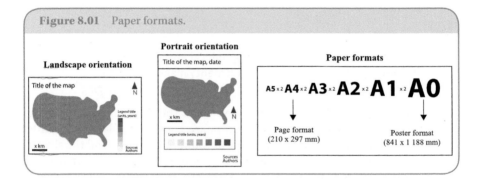

**Figure 8.01**    Paper formats.

## 8.1 "DRESSING" THE GEOGRAPHICAL IMAGE

The finalization of the map is manual for a large part. Most of the time computer-aided design (CAD) software is used. The first stage is to "dress" the geographical image that has been developed. In fact, once decision has been made about projection, basemap, and symbolization, the composition of map elements can be made. Certain elements are essential (title, legend, source, scale, dates), and others are optional (orientation, inset maps, toponyms, or other texts, frame, graticule, drawings, etc.). These will be detailed in the following figure.

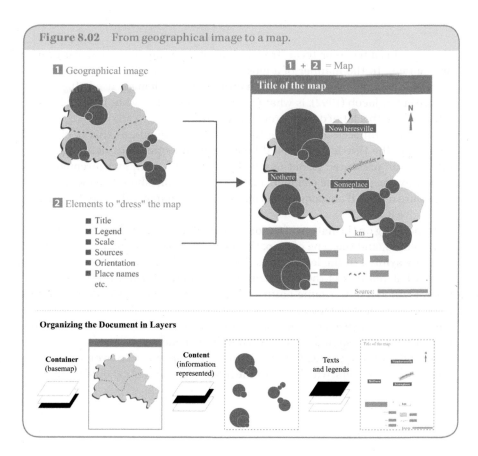

**Figure 8.02**    From geographical image to a map.

1 Geographical image

1 + 2 = Map

Title of the map

N

Nowheresville

Dotted border

Nothere

Someplace

2 Elements to "dress" the map

- Title
- Legend
- Scale
- Sources
- Orientation
- Place names
  etc.

km

Source:

**Organizing the Document in Layers**

**Container**
(basemap)

**Content**
(information
represented)

**Texts**
and legends

Title of the map

## 8.1.1 Title

This is an essential element for the map and one of the points of entry for the message to be delivered. It is also what is seen first. It should be brief and in sufficiently large print to stand out. It can be eye-catching, dull, discreet, openly militant, or, conversely, neutral and merely descriptive. In all events, the title conditions the way in which the map will be read. It carries an implicit commentary on the image offered. In other words, it provides an angle and a point of view. Drafting it is therefore an important step, and a map always needs a title.

### 8.1.2 Legend

The legend is a term-for-term "translation" of the graphic elements appearing on the map. In the manner of a glossary, it is the interface between the cartographic image and the precise meaning of each element used. The legend, according to Jacob (1992), is what is read as opposed to what is seen. Legends often deal with a lot of design problems as for maps.

First, it's unnecessary to title the legend as "legend", as this is obviously what it is. The title legend should be as explicit as possible, detailing all the information presented (exact name of the variable, year, measurement unit). Generally speaking, the elements making up the map should be included only if they are useful. If an element has no use (a place name, a line, etc.), it is better to remove it.

Depending on the nature of the data presented, the form in which the image is presented, and the type of map, the legend is designed differently. Be aware that any symbol in the legend must look exactly like the symbol on the map. Also, what matters is the organization and hierarchization of the elements used. Elements of the same nature or on the same theme should be grouped to facilitate comprehension. Do not hesitate to provide details on the type of data and the measurement units used, for instance, by way of a legend sub-title. Numerical values should be presented in homogeneous manner (coma for the thousands, values rounded off, etc.), and it is important to avoid using symbols such as brackets, which clutter the image. All elements should be carefully aligned.

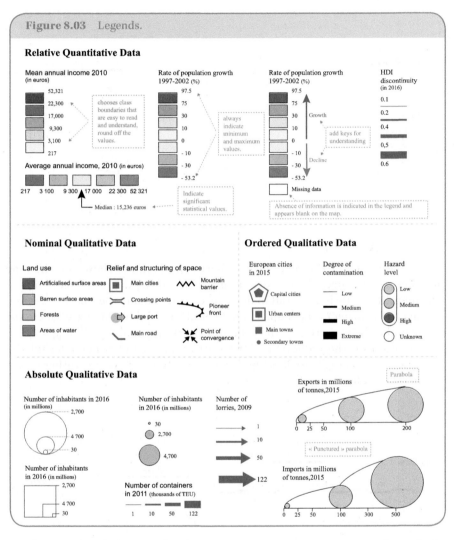

**Figure 8.03**    Legends.

The graphic balance of a legend is just as important as that of the cartographic image.

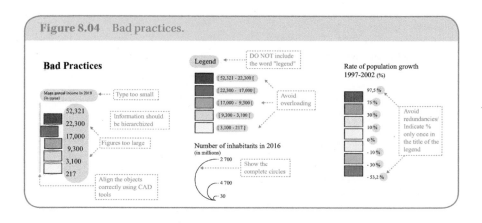

Figure 8.04    Bad practices.

## 8.1.3 Scale

The scale indicates the relationships between distances shown on the map and actual distances. It provides information on the degree of reduction of the map. The scale is essential in most cases, but it should be small and unobtrusive, in a corner of the map. It can be omitted from world maps (where the scale will only be correct in the center of the projection) and on maps where information on the reduction of the space represented is not relevant (for instance, in a cartogram). It is highly advisable to use a graphic rather than a numerical scale.

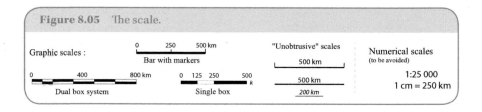

Figure 8.05    The scale.

### Definition

**Small and large scale:** A scale is said to be "small" when the space represented is large, and the scale is said to be "large" when the space represented is small. In geography, we therefore refer to a large scale not in relation to the world, but on the contrary to talk about a local situation. While this may appear counter-intuitive, it is nevertheless mathematically sound: a scale of 1/25,000 is "larger" than a scale of 1/500,000.

● **FOCUS: Maps on Screen and Maps on Paper**

"In that Empire, the Art of Cartography attained such Perfection that the map of a single Province occupied the entirety of a City, and the map of the Empire, the entirety of a Province. In time, those Unconscionable Maps no longer satisfied, and the Cartographers Guilds struck a Map of the Empire whose size was that of the Empire, and which coincided point for point with it. The following Generations, who were not so fond of the Study of Cartography as their Forebears had been, saw that that vast Map was Useless, and not without some Pitilessness was it, that they delivered it up to the Inclemencies of Sun and Winters" (Borges, 1982).

Beyond its humoristic dimension, this passage from Borges focuses on the role of the map as a simplification of the real world. A map representing the world as a whole, besides being unwieldy, is also useless. Since it does not resort to abstraction, a map of this sort would miss its main function: modeling geographic space to makes it apprehendable. According to Nelson Goodman, if maps become as large and, in all respects, identical to the territory that is mapped – and indeed well before that – the aims of a map are no longer met. There is no map that is completely adequate, because inadequacy is inherent in maps (Goodman, 1972). Yet the myth of the map on a scale of 1:1 remains. Today, it is digital. Google Maps, Open Street Map, or Bing digitize the world, its details, and its every corner. Would the Borges map of a globalized empire help us to know more about the way the world works? It is highly unlikely.

## 8.1.4 Sources

Sources must appear on the map. This concerns the sources of the data, which should be scrupulously detailed (complete name, date of the data, date of download, etc.) and the source of the map itself: name of the author and any indications concerning copyright, licensing, etc. The author should take responsibility for his or her map by signing it. However, when there are several authors in a collective enterprise, the names of the authors can be omitted and replaced by the name of the collective (association, firm, organization, etc.). A map without its sources cannot be checked or contested. It is therefore not reliable.

● **FOCUS: The Creative Commons License**

The Creative Commons (CC) license is a system of flexible licenses enabling the authors of maps to define the way in which their cartographic productions can be diffused and re-used: attributions, banning of commercial use, sharing under the same conditions, no alterations allowed. Based on the model of free licenses, this system is particularly well suited to the diffusion of maps on the internet. ●http://creativecommons.fr/licences

### 8.1.5 Dates

Dates should always appear on maps. There are several types – the date of the data used (in some cases the date of the download), the date of the indicator represented, and the date of the actual production of the map. Sometimes, the date attached to the data can be given as the year (population in 1999), or by a precise date (population on January 1, 1999) or by an interval (number of births from January 1, 1999 to December 31, 1999).

### 8.1.6 Orientation

The orientation of the map is of use when the space represented is small, not clear, or not oriented towards the north (since the north is the default orientation when maps are indeed oriented towards the north). Conversely, for thematic maps on a small scale (America, the world), it is perfectly acceptable not to show the orientation, since this information does not contribute to comprehension. In addition, on these maps the representation of the north can be misleading as a result of the projection used.

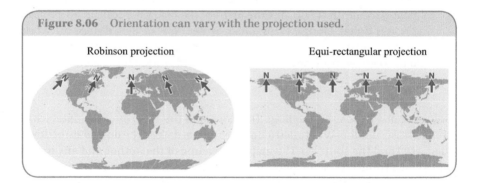

**Figure 8.06**    Orientation can vary with the projection used.

Robinson projection                Equi-rectangular projection

● **FOCUS: Did You Know?**

The word orientation is derived from the Latin word *oriens*, meaning the east, i.e., the direction of the rising sun and the Holy Land (while the west was associated with the finitude of the uncrossable Atlantic Ocean). Thus, etymologically speaking, orienting a map to the north is not logical!

### 8.1.7 Place Names

Place names or toponyms can be very helpful for the reading and understanding of the map. The choice of these place names is not innocent. It consists in representing and thus focusing on place names deemed to be important in relation to the theme of the map or its message. The information retained thus needs to be selected and hierarchized. In addition, certain places can be the subject of dispute, and there may not be a consensus on their naming. This means the cartographer has to take sides.

There are many options for positioning and presenting these different "text" objects – the font, upper or lower case, italics, bold, etc. When capital letters are used, accents should be placed where needed. It is useful to note that while the use of different types of print was fairly codified on older maps, there are no rules on this matter for contemporary theme maps. Place names enable enhanced precision or a focus on certain places, but once again it is important not to hinder the perception of the image, but rather to favor it. For instance, you can choose to show the names of towns and cities in lower-case print, the names of countries in capital letters, and the names of water courses in italics. These different types of print (and possibly different colors) enable a differentiation and a hierarchization of the different types of geographical objects represented.

It is likewise important to organize the graphic document by grouping all the text in a single layer so as to allow for any later processing (translation, changes of font, etc.).

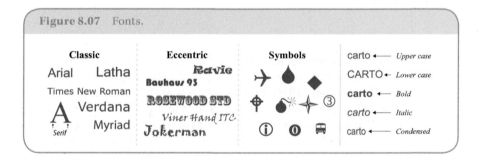

Figure 8.07    Fonts.

## 8.1.8 Insets

Insets are small maps used to complement the main map. They can zoom onto a part of the map to make it more legible or show the broader localization of the space represented.

## 8.1.9 Supplemental Text

Supplemental text can be added. Some elements must be on the map (source statement, for example), whereas others could be optional. You could add a small framed inset beside or within the cartographic image containing different types of information on the map: the author, technical elements, sources, or any other elements required. Be careful not to include too much elements in this frame which will lose the maps.

### 8.1.10 Any Other Relevant Elements

Other graphic elements can be added to the map – graticule, the frame, or any other layer serving to "dress" the map and make it clearer and more readable, or to help the user find his or her bearings (forests, water courses, large cities, etc.).

## 8.2 LAYOUT AND STAGING

Once the map has been "dressed", it still needs to be finalized. This generally means choosing the layout on the page. The task is to arrange the different elements to be used to "dress" the map around the geographical image, minimizing blank spaces, so as to render the overall effect intelligible. The different elements need to be positioned hierarchically so as to show what is important and what is less so.

● **FOCUS: The Cartographic "Stage"**

The idea of the cartographic "stage" is a borrowing from the cartographer Philippe Rekacewicz, who sees the map as a theater stage with its sets and its actors (Rekacewicz, 2014). Staging, which is envisaged as the art of staging the action in the theatrical world, is indeed a notion that fits mapping very well. We can also recall the first atlases by Abraham Ortelius in the 16th century, named "theaters of the world", and also the world maps of the Middle Ages sometimes referred to as "estoire" (history or story).

The layout on the page can also be the result of a more elaborate process, which can also be called staging. Not simply an operation consisting in placing the different elements, staging means that the map becomes an object that genuinely transfers knowledge. It is not just a technical operation; it is the phase during which the graphic elements are set out in such a way as to show the coherence of one with another, so as to create meaning. As in theatrics, staging means designing the stage setting and organizing the way in which the narrative is to unfold.

Creating a map thus means telling a story based on evidences. To do this, you need to choose the right words and the right colors, to hierarchize the information, and guide the user step by step so as to deliver the chosen message in effective manner. The staging of the map makes it possible to construct trajectories for its perusal. By way of the choice of toponyms, making certain elements less conspicuous and others more obvious, the author of the map can catch the viewer's eye so as to skillfully deliver the intended message. It is at this stage in the production of a map that the cartographer will decide whether or not to be implicated in the message delivered. In addition, the message will be envisaged differently according to the user that is targeted. If the map is to deliver an intelligible, straightforward, spatial message, straightforwardness and intelligibility are conceived differently according to the person or group of persons to be targeted. Thus, the question of the "audience" is a prime, determining criterion in the very conception of a map. It will not be constructed in the same manner for a newspaper, a political decision-maker or a scientific report. The words chosen, and the implicit elements, will differ.

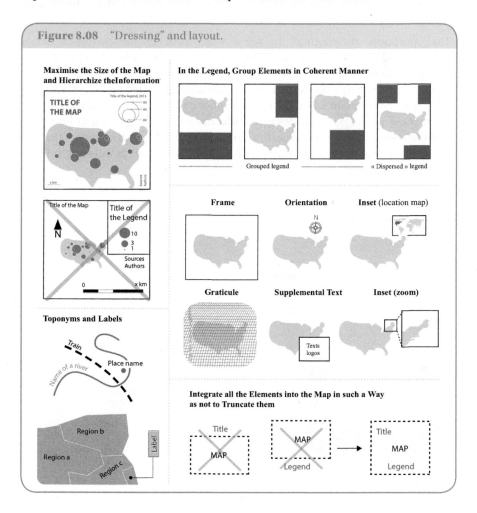

**Figure 8.08** "Dressing" and layout.

● **FOCUS: An Effective Map Is a Pretty Map**

The As in a painting, the lines and patches of color on a map lend themselves to artistic expression. Maps are often beautiful, stylish, and attractive, and this is an asset. However, we can also regularly see that many maps are unattractive. They are maps where the aesthetic dimension has been set aside in favor of scientific or technical constraints. Indeed, it is often considered that, for scientific reasons, superfluous elements that do not directly serve the purpose of the map should be removed. Ultimately, these maps without their frills are dull and insipid, and not very effective. An effective map is attractive, and it draws the eye and captivates. It is a map you want to look at. So do not be content with an ugly map, produce a work of art!!

The operation of "staging", in some ways akin to the novelist's art, is complicated. While graphic semiology can readily be understood by geography students because it concerns technical aspects, mastering the narrative of staging is a lengthy process. It requires knowledge, skills, and a lot of experience. It is the moment when the cartographer can set out and defend his or her view. The drafting of a well-designed title, the focus on a particular boundary, and the choice of an original cartographic projection are all elements that can radically change the way a map is approached. Thus, there is only one rule: use your imagination!

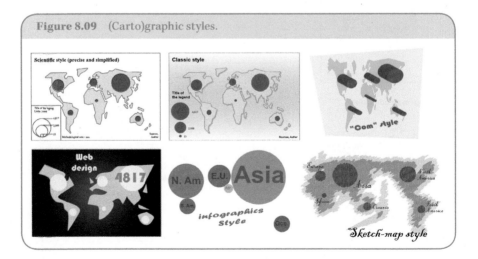

**Figure 8.09**    (Carto)graphic styles.

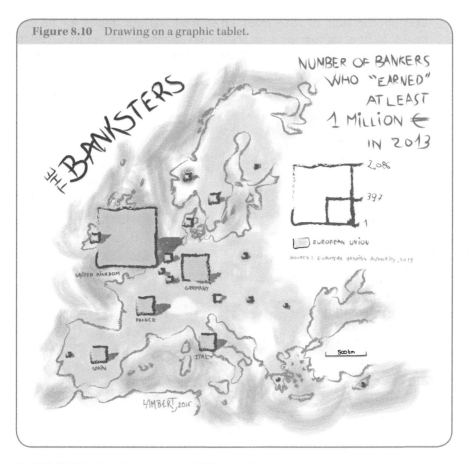

**Figure 8.10** Drawing on a graphic tablet.

● **FOCUS: GeoArtist or cARTographer?**

The proximity between art and cartography is long-standing, and the two disciplines have in some cases been united in a single person. Leonardo da Vinci was at once an engineer, a scientist, a painter, and a cartographer. This proximity can still be observed today. Many cartographers have made their maps genuinely artistic objects. Cartographic expression is in many ways, a form of artistic expression – touch-maps painted or sketched maps, sculptured maps, sound maps, etc.. On the other side, many contemporary artists also use geographical maps as the base or as material for their creations. Thus, maps can be cut up, used in collages, or remodeled to contribute to a creation. For instance, there is Michel Houellebecq's book *The Map and the Territory* where the main character uses photographs of road maps taken from various angles for his creations. In fine, cartographic or artistic creations can be realistic, seeking to represent reality as far as possible, or they may draw away from reality and seek to communicate an impression, a viewpoint or an idea (as is the case with cartograms).

**Figure 8.11**    Improving graphics for map readability.

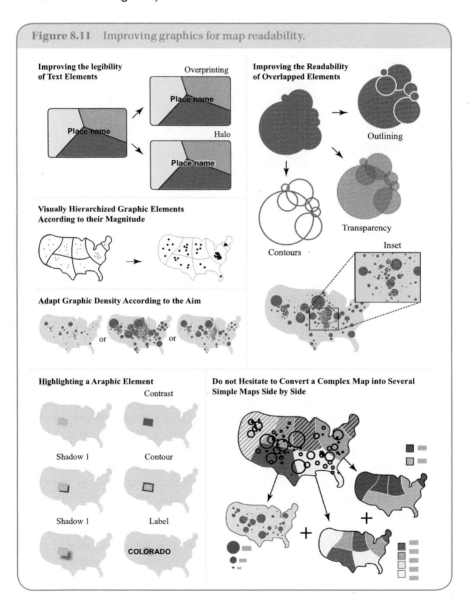

**Figure 8.12**    A process of "staging" a map.

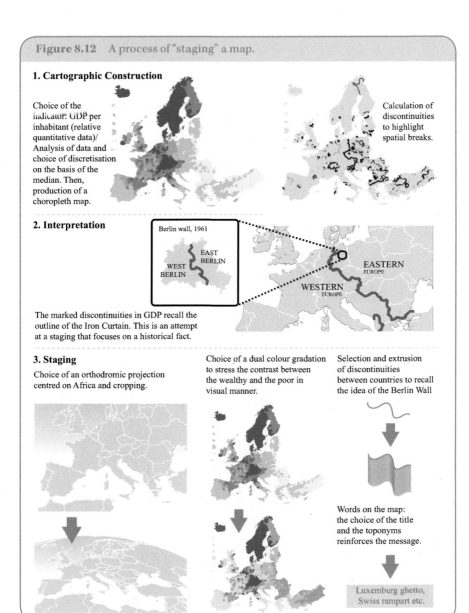

**1. Cartographic Construction**

Choice of the indicator: GDP per inhabitant (relative quantitative data)/ Analysis of data and choice of discretisation on the basis of the median. Then, production of a choropleth map.

Calculation of discontinuities to highlight spatial breaks.

**2. Interpretation**

Berlin wall, 1961

WEST BERLIN    EAST BERLIN

EASTERN EUROPE

WESTERN EUROPE

The marked discontinuities in GDP recall the outline of the Iron Curtain. This is an attempt at a staging that focuses on a historical fact.

**3. Staging**

Choice of an orthodromic projection centred on Africa and cropping.

Choice of a dual colour gradation to stress the contrast between the wealthy and the poor in visual manner.

Selection and extrusion of discontinuities between countries to recall the idea of the Berlin Wall

Words on the map: the choice of the title and the toponyms reinforces the message.

Luxemburg ghetto, Swiss rampart etc.

**Figure 8.13**   Example of the staging of a map.

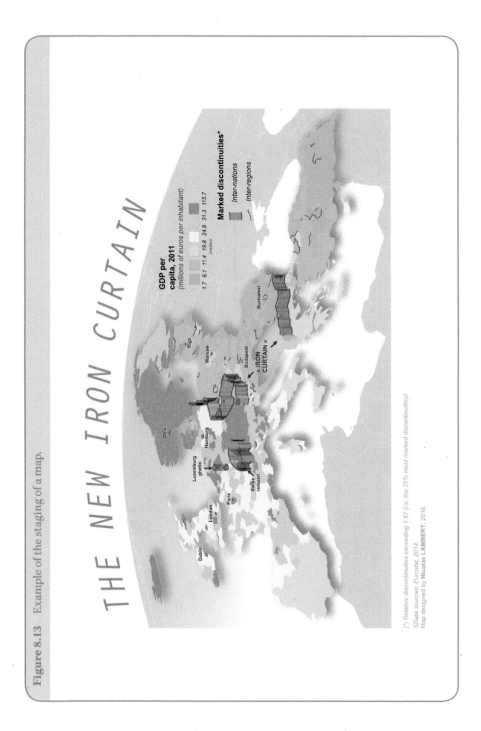

THE NEW IRON CURTAIN

GDP per
capita, 2011
(millions of euros per inhabitant)

1.7   6.1   11.4   19.8   24.8   31.3   115.7
(median)

Marked discontinuities*

Inter-nations
Inter-regions

«IRON
CURTAIN»

Dublin
London
Paris
Luxemburg
ghetto
Swiss
rempart
Hamburg
Oslo
Warsaw
Riga
Budapest
Bucharest

(*) Relative discontinuities exceeding 1.87 (i.e. the 25% most marked discontinuities)
SData sources: Eurostat, 2014.
Map designed by **Nicolas LAMBERT**, 2016.

*"From Stettin on the Baltic to Trieste on the Adriatic, an iron curtain has fallen on the continent" (Winston Churchill, 1946). Europe is divided into two: the rich to the west and the poor to the east. Between the two, an immaterial wall has been erected. Except for former Eastern Germany, this "wall" closely follows the boundary of the Iron Curtain, despite the fact that it fell in 1989. It was formerly materialized by barbed wire, mine fields, and watch towers, but the new Iron Curtain today is economic.*

*Beyond the contrast between East and West, this map shows us a lot about what is happening in the West. While the heart of Europe and northern Europe are undoubtedly to be placed among the wealthy regions (in orange on the map), the regions of southern Europe (in light green) are characterized by a per capita gross domestic product (GDP) that is below the European median. This means that half the European regions are wealthier than they are. Conversely, in addition to the prosperous urban concentrations that can be seen on the map (Paris, London, Dublin, Hamburg, etc.), wealthy Western Europe has extremely wealthy enclaves on its territory (Luxemburg, Switzerland).*

## 8.3 DESIGNING A POSTER OR A CARTOGRAPHIC COMPOSITION

A cartographic composition can be designed for a paper format or for a digital format. It is a document that comprises several graphic and/or cartographic elements coherent with one another. This may concern a single phenomenon represented on several geographical scales or the juxtaposition of several illustrations of the same theme. The graphic elements should be comparable one to the other as far as possible (same discretization, comparable proportional symbols, etc.). The theme should be clearly identified. While each map making up the overall composition can have its own title, a general title is always needed to recall the coherence of the different elements.

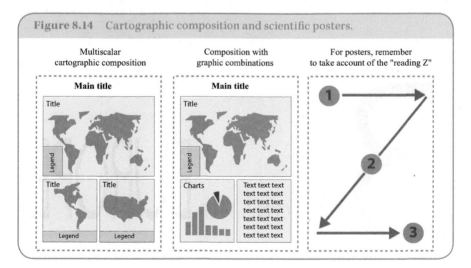

**Figure 8.14**   Cartographic composition and scientific posters.

A scientific poster is a cartographic composition where the general layout is designed to guide the eye of the viewer from one element to the next. Here, it is not a simple juxtaposition of maps that are used, as in the case of a cartographic composition. A poster always has a theme to discuss, and the graphic and text elements provide answers. It is important not to seek to include too many elements of information at the same time. It is better to deliver a few well-explained elements than dozens that will not be retained. Quantity will not impress the viewer.

**Figure 8.15**    Cartographic composition about sustainable development.

The size of the different objects is defined according to their importance. The viewer is guided visually from one element to another by way of a graphic strategy. In matters of communication, it is often said (for Western cultures) that the eye scans following a Z. It starts at the top left and then travels across

**Figure 8.16**   Scientific poster on the theme of migration.

the image in the usual reading direction. Thus, the design of a poster has a lot to gain from following this logic, highlighting certain key elements: the eye is attracted by images, bright colors, words in bold print, etc.

The elements that will catch the attention of the viewer the most are first of all images in color (photographs, drawings, maps), then diagrams and tables, and finally text elements. The poster is a visual tool, so that the emphasis should be on figures, graphic elements, images, and maps, limiting text in favor of illustrations. Any text should be very brief; use simple words and be correctly articulated with the images. Maps that were designed in other settings should not be pasted directly into a poster without first of all reviewing the general staging (changing colors, moving legends, resizing place names, etc.).

Posters are generally designed in A0 format (841 × 1,189 mm), and it should be possible to read the information from one meter away or more (in particular in case of crowds or for the many viewers whose sight from a distance is poor).

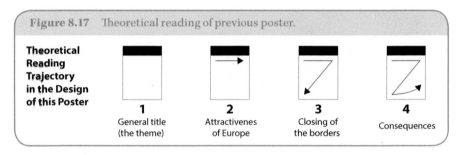

**Figure 8.17**    Theoretical reading of previous poster.

## Quiz

- How can a map be made more readable?
- What are the essential elements to "dress" a map?
- What is an inset?
- What could be called supplemental elements?
- What is a map "on a large scale"?
- How is an effective cartographic composition designed?
- Should a poster be constructed in the same way as a cartographic composition?

# Chapter 9

# Variations on a Theme

- Knowing and understanding cartographic subjectivity
- Developing a critical approach
- Learning to construct a cartographic story.

The purpose of this chapter is twofold: it has a practical objective, enabling the student cartographer to follow the process of constructing a map step by step, and a theoretical objective aiming, by way of examples, to show what cartographic subjectivity actually is in concrete terms. Designing a map is always a narrative.

● **FOCUS: Variations on a Theme**

In 1947, Raymond Queneau published *Exercises in Style* in which he produces the same narrative 99 times in 99 different ways. In mapmaking, the *styling exercise* (or variations on a theme as in music) is also possible: a given data can be "narrated" or "told" in many different ways.

## 9.1 WHAT IS GDP?

Since Adam Smith (1776), the question of the measurement of wealth has been the subject of considerable debate among economists. Gradually, one indicator (and one viewpoint) has come to dominate the scene: gross domestic product (GDP). No other indicator is more widespread in the media, alongside its well-known off-shoot, growth rate.

GDP measures the level of production of a country in the course of one year. GDP per inhabitant (or per capita GDP) is often *used as* a quality-of-life indicator. GDP, invented in 1934 by Simon Kuznets, came into its own after the World War II, at the time when the countries of Europe were reconstructing.

It is generally viewed as a good indicator of economic activity. The indicator does nevertheless have numerous limitations.

● **FOCUS: What GDP Does Not Measure?**

1. GDP does not take home-consumed production or voluntary work into account. Growing apples in an orchard and consuming them in the home does not enter into the calculation of GDP. However, selling these same apples on markets contributes to GDP. The same is true for work in the home: household chores are not taken into account except if they are performed by a salaried employee. Generally speaking, non-commercial activities do not enter into the calculation of GDP.

2. GDP does not take account of externalities. If a business produces waste, emits $CO_2$, or pollutes the aquifer, this has no negative impact on GDP (at least in the short term). Further to this, when the need arises to create activities to depollute or process waste, they will increase the GDP. When one builds, the GDP rises, when one destroys, it rises, when one pollutes, it rises, and when one depollutes, it rises – magic!

3. GDP provides a measure of production, without providing information on the content of this production. Producing flowers, weapons, or toxic substances all have the same impact on GDP. It is thus possible to increase GDP, as suggested by Keynes in 1936, by digging holes in the morning and filling them in the same evening. Absurd!

4. The GDP of a country measures production, but in no way how the production is distributed across the population and across the territory. Whether there are a lot of wealthy people alongside a lot of poor people, or just a lot of middle-class people, has no direct impact on GDP.

While this indicator is open to debate for several reasons, its cartographic transcription can take on numerous forms. This chapter will explore this scope.

## 9.2 HOW CAN GDP BE MAPPED?

While the interpretation of GDP requires caution, mapping this value is eloquent. In this chapter, we propose "variations on the theme" aiming to show how a single statistical indicator can be mapped in different ways to produce messages that are different or even contradictory. The material we have to hand is as follows: a basemap of the countries in the world (where it is possible to alter the projection) and a table giving the GDP in dollars and the population of each country in 2014. From this single set of objective data, we will construct eight cartographic representations, each with its own message.

**Map 1** provides a consensual picture of the world where each country is colored according to its level of wealth per inhabitant, i.e., in dark blue, the wealthiest countries and in light blue, the least wealthy. The simple gradation of blues enables a classification of countries according to their level of prosperity. The title and the legend are purely descriptive. The title points to a world where each county draws advantage from globalization, each growing at its own pace. This deceptively neutral map is based on the Mercator projection, which besides the fact that it is widely found in classrooms has the characteristic of over-representing the countries in the northern hemisphere, because it does not reflect the actual size of the different countries. On this map, it is the wealthy countries in the north that are highlighted. This map forms geographical proof of the supremacy of the wealthy countries in the north over the world order.

**Map 2** provides a very different picture. It is based on an upturned Peters projection (the south is at the top), and it encourages the viewer to change his or her outlook on the world. This is also the title of the map. This projection is said to be "equivalent" – it reflects the actual relative surface areas across countries. This map, widely welcomed by alternative globalization proponents, puts the emphasis on countries in the south. This aspect is reinforced by the upturning of the basemap, which, as well as disorienting the viewer, places these countries at the top of the map. To stage a world where we find the rich to one side and the poor to another, the discretization method used is based on the median.

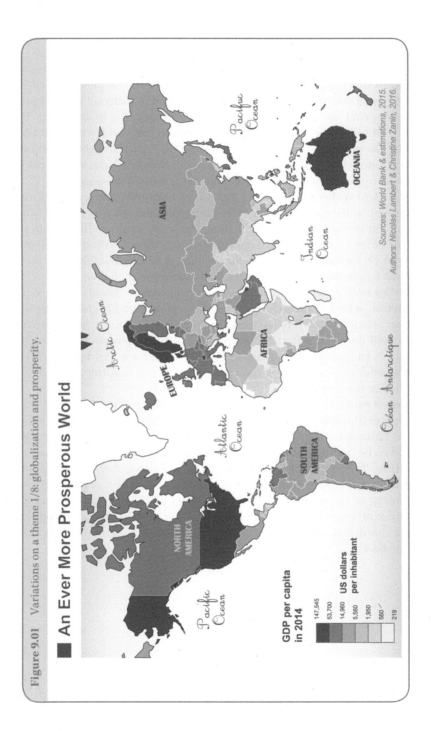

**Figure 9.01**  Variations on a theme 1/8: globalization and prosperity.

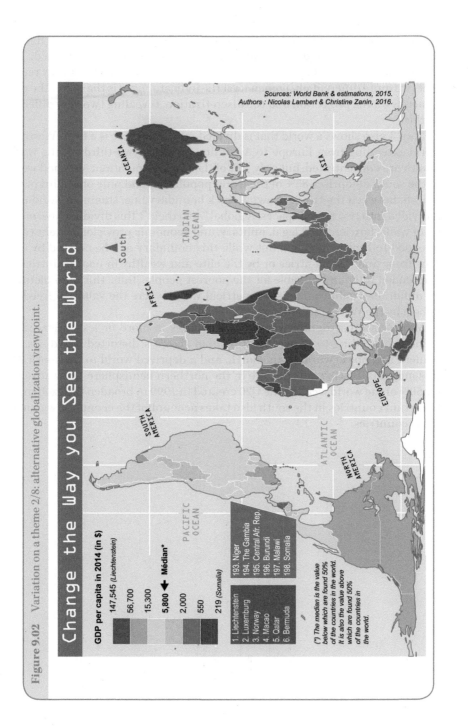

Figure 9.02 Variation on a theme 2/8: alternative globalization viewpoint.

The contrast is reflected graphically by a dual color gradation. The six wealthiest countries and the six poorest countries are specifically named.

The comparison of maps 1 and 2 could give the impression that they represent two quite different phenomena, yet the indicator used is the same. The play is merely on the projection and the discretization, revealing two very different worlds.

**Map 3** also shows a world that is divided into two. There is a first "Western" world extending from Europe to Australia, in which two-thirds of the world wealth are concentrated, but only 15% of the world population. A second world to the south comprises 85% of the world population and only one-third of the wealth. Between the two is drawn a thick boundary line, staging a world that is divided into two closed spaces. A global "apartheid". This dividing line, made up of walls, barriers, and legal, military, and economic provisions, is above all intended to hinder mobility. But while this boundary can be crossed by citizens from wealthy countries or by the elite and wealthy in poor countries, it is intended to be uncrossable for the poorest people. Thus, this map pictures an unfair, inequalitarian, dissymmetrical world where the values of freedom upheld by the West are challenged.

**Map 4** is a cartogram where each country is more or less enlarged or reduced according to its wealth. This map shows a distorted world, i.e., an opulent and obese world to the north, and a deprived world to the south. On this map, all the G8 countries are in the northern hemisphere. They account for 50% of the world wealth. The G20 created in 1999 to broaden the exchanges to certain countries in the south also takes account of the economic weight of these countries.

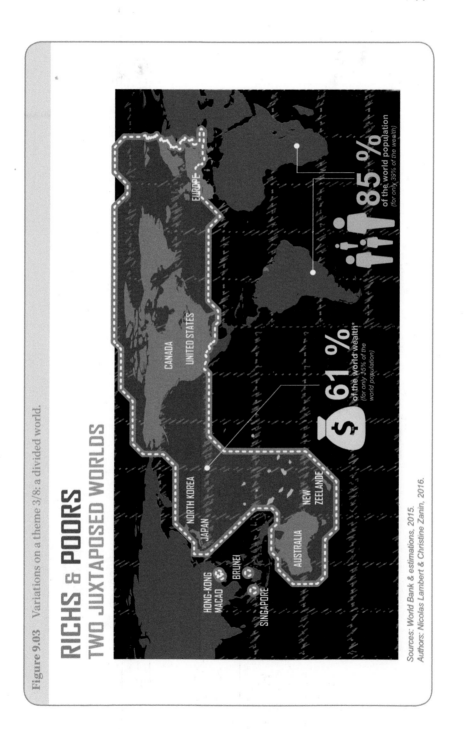

**Figure 9.03** Variations on a theme 3/8: a divided world.

RICHS & POORS
TWO JUXTAPOSED WORLDS

61 %
of the world wealth*
*(for only 35% of the
world population)*

85 %
of the world population
*(for only 39% of the wealth)*

CANADA
UNITED STATES

EUROPE

NORTH KOREA
JAPAN

HONG-KONG
MACAO
BRUNEI
SINGAPORE

AUSTRALIA
NEW
ZEELANDE

*Sources: World Bank & estimations, 2015.*
*Authors: Nicolas Lambert & Christine Zanin, 2016.*

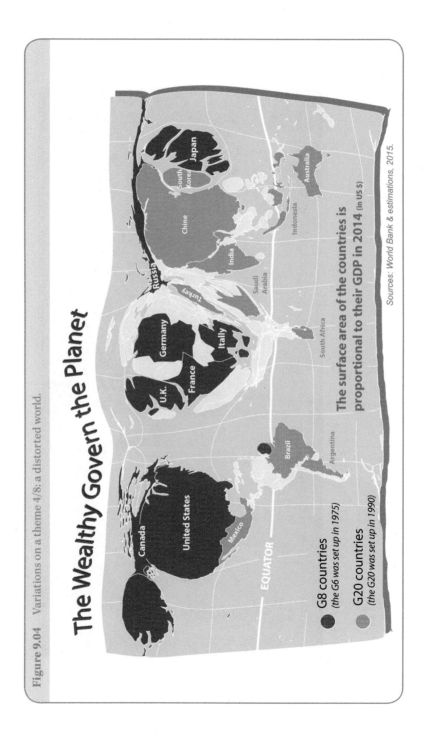

**Figure 9.04**  Variations on a theme 4/8: a distorted world.

The 19 countries making up the G20 account for 77% of the world wealth (the EU is the 20th member). Ultimately on the strength of their wealth and economic weight on the international scene, just a handful of countries decide for the rest of the world. They dominate it, govern it, and organize it. The map likewise shows the countries that are forgotten. Reduced to almost nothing on the map, they have no say. This map reduces their territories to virtually nothing. They are the discounted in a globalized world. Yet they account for more than 40% of the world population.

**Map 5** uses the same information as Map 4, which is absolute quantitative data. Here, it is transcribed graphically by the visual variable "size" represented by proportional squares positioned in the centers of the countries. As this type of cartographic representation is closely linked to the underlying grid and to the size of the territorial units, it highlights the small size of the European countries. In Europe, the squares are superimposed, suggesting a wealthy but disunited continent, while at the same time, the BRICS countries (Brazil, Russia, India, China, and South Africa) appear as powerful, homogeneous blocks. This map thus points to a divided, weakened Europe faced with emerging nations with increasing weight on the international scene. This map is in fact a plea for a pro-European view, for a union of European countries, the only way to confront international competition.

**Figure 9.05**    Variations on a theme 5/8: a pro-European view.

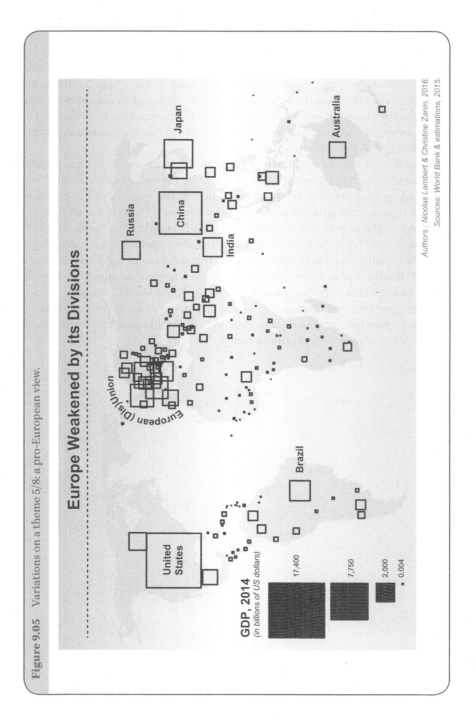

Europe Weakened by its Divisions

Russia

Japan

China

India

Australia

European (Dis)Union

United States

Brazil

GDP, 2014
*(in billions of US dollars)*

17,400

7,750

2,000

0.004

*Authors : Nicolas Lambert & Christine Zanin, 2016.*
*Sources: World Bank & estimations, 2015.*

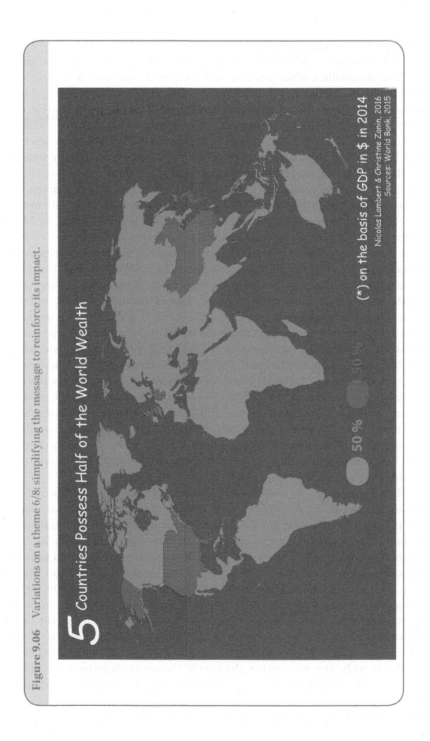

**Figure 9.06**  Variations on a theme 6/8: simplifying the message to reinforce its impact.

**Map 6** provides a highly simplified representation of the distribution of wealth across the world. The GDP variable is discretized into two classes by the quantile method. The median, which is the value that separates the two classes, has a value of 32,830 million dollars. Only four countries are above this value. Thus, this map showing two classes tells us that four countries (USA, China, Japan, and Germany) possess as much wealth as all the other countries in the world. With a procedure of this sort, what the map loses in detail it gains in clarity. The information is indeed highly caricatured, but the resulting message and the simple image have an impact.

**Map 7** is an allegory of class warfare and the collapse of nation states in the face of the multinationals. In 2005, the American billionaire Warren Buffet stated in a television broadcast: "There's class warfare, all right, but it's my class, the rich class, that's making war, and we're winning" (interview on *CNN*, May 25, 2005). This map is an illustration of his words. On this image, certain countries have disappeared from the face of the Earth. They are the rich countries. As they do not serve the purpose of the map, they have been simply erased. The remaining countries are the poor countries. New continental outlines appear, made up of islands and islets scattered over the surface of the globe. Poorest among the poor, one continent appears massive – the African continent. The world seen here is not Warren Buffet's world, but the world of one of the wealthiest men in the world, Bill Gates. In this world, there are 132 countries, each individually less wealthy than the multibillionaire. The title of the map suggests a question: who is less wealthy than Bill Gates? These countries overall, or each one separately?

While the construction of this map can be disputed on scientific grounds – it compares stock data (the wealth of Bill Gates) with flow data (the quantity of wealth produced in one year by a country) – the message is intentionally controversial. Class warfare or disputes between nation states and multinationals? – This map delivers a simple, clear message: the world is now dominated by the ultra-rich. It is an image to generate reflection.

**Map 8** sets out to be optimistic. While it highlights inequalities in wealth across the world, it advocates sharing. The cartographic projection chosen is the polar projection, so that the map can be viewed from all directions. In the center of the map there is nothing, no particular country is highlighted. Each country is positioned on an equal footing. It is an image of a unified world, like the cartographic projection that is part of the United Nations logo. The variable represented results from the conversion of the outset data. On the basis of logic of equal distribution, the indicator measures the quantity of wealth to be redistributed for each country to have the same per capita GDP. Certain countries, in red, (European countries, USA, Japan, etc.) thus need to redistribute wealth. Other countries, in green (India, African nations, China) need to receive wealth. By construction, the sum of the amounts given and received is balanced. This map is naive, a "nice" view of things. It is a call for much greater international solidarity, rather unlikely to occur. But even if it is not very realistic, its plea is for a mutually supportive view of the world, as in its title.

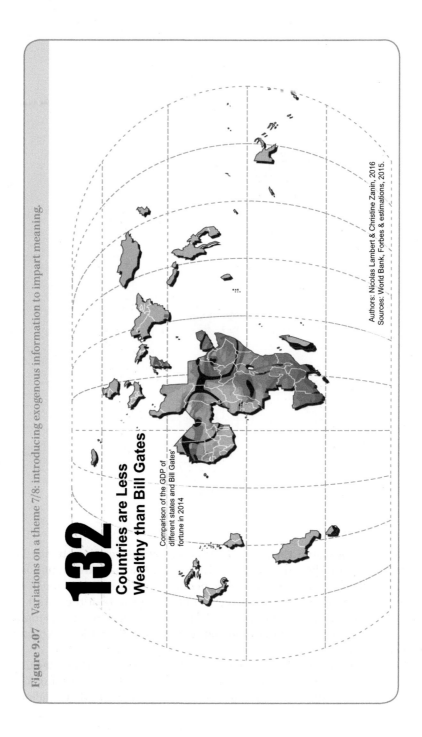

**132**

**Countries are Less Wealthy than Bill Gates**

Comparison of the GDP of different states and Bill Gates' fortune in 2014

Authors: Nicolas Lambert & Christine Zanin, 2016
Sources: World Bank, Forbes & estimations, 2015.

**Figure 9.07** Variations on a theme 7/8: introducing exogenous information to impart meaning.

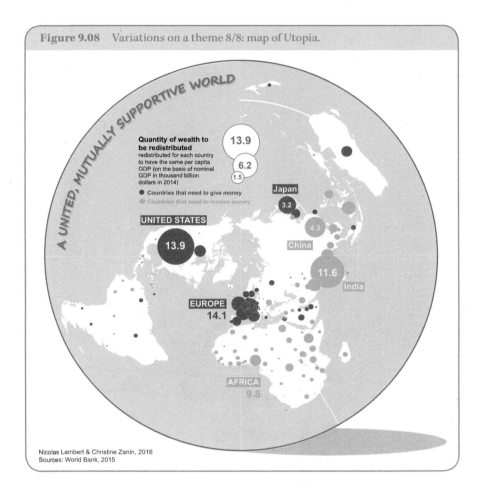

Figure 9.08    Variations on a theme 8/8: map of Utopia.

A UNITED, MUTUALLY SUPPORTIVE WORLD

**Quantity of wealth to be redistributed**
redistributed for each country to have the same per capita GDP (on the basis of nominal GDP in thousand billion dollars in 2014)

13.9
6.2
1.5

● Countries that need to give money
● Countries that need to receive money

UNITED STATES
13.9

Japan
3.2

China
4.3

11.6
India

EUROPE
14.1

AFRICA
9.8

Nicolas Lambert & Christine Zanin, 2016
Sources: World Bank, 2015

## 9.3  WHAT LESSONS SHOULD BE DRAWN?

These variations on a theme (or *Exercises in style* according to the French author Queneau) show that facts can be narrated in many different manners. Facts, without any interpretation or perspective, do not have much to say. Here, all the maps were constructed from the same data at the outset – GDP and population. But the message delivered is the work of the cartographer who offers his or her interpretation and cues by way of the title and the graphic choices made.

Ultimately, which is the best map? Providing the construction rules are complied with, all are fair representations, variably relevant and useful. Yet they suggest views of the world that are very different: a world that is sometimes harmonious, sometimes divided between rich and poor, sometimes mutually supportive, and sometime ruthless. The process of construction from the raw data to the theme map is thus both a technical process and a rhetorical process. Without a specific viewpoint, reality has nothing to say.

 **Quiz**

- What is GDP?
- What is the importance of the narrative of a map?
- Can several different maps be derived from the same statistical data?

# Conclusion: Cartography Is a Martial Art

Geography maps are today part of our daily lives. We see them on television, in newspapers, and on the internet – they describe facts, they explain the facts, and they help to understand them. By portraying arguments, they also serve to defend or contest this or that view of the world. Indeed, the complexity of the world cannot be reduced to a consensual representation of some sort that could be accepted by all. To be useful and to serve to back up argument in a debate, maps need to be plural. The confrontation of maps makes us reflect and enables each of us to form an opinion. In fact, to be useful, a map needs this plurality and even a degree of controversy.

● **FOCUS: John Brian Harley and the "Power of Maps"**

John Brian Harley (1932–1991) was a British cartographer specialized in the history of cartography. He considered the map to be a tool enabling a government to exercise control and power over a territory, and he was the first to characterize the map as a social construction rather than as an objective representation of reality. According to Harley, every map carries an ideological discourse, which is none other than the discourse of dominant individuals imposing their values and beliefs on society. In a series of ground-breaking articles, he developed the theory of the "power of maps" and had a central role in the emergence of critical cartography in geography.

One of the conditions required for an honest cartographic debate is transparency. If a map, as we consider, is indeed a particular view of the world, the way in which each particular message is constructed should be traceable and decipherable, to fuel a rational, argued debate. To ensure this necessary traceability, each source needs to be detailed, notes should be provided to describe the methods used, and the data and the basemap used should be identified as freely as possible.

Committed cartography, which is what we advocate here, is thus part of another battle: the freeing of data (open data) and access to open-source software. Any map should thus be open to being challenged. Any representation should be considered provisional; it should be open to discussion and to contestation.

● FOCUS: Cartography Seen as a "Fighting Sport"

This idea is a reference to the French sociologist Pierre Bourdieu (1930–2002). Considered as one of the most important sociologists of the second half of the 20th century, his thinking had a great influence in humanities and social sciences. Pierre Bourdieu was also known for his public commitment. From his point of view, sociology was a weapon to deconstruct mechanisms of domination. In 2001, a French documentary film directed by Pierre Carles, entitled *La sociologie est un sport de combat* (Sociology is a fighting sport) was dedicated to him. It is this lineage that we are taking up again here.

In his essay *Indignez-vous* (translated as "Time for outrage"), Stéphane Hessel wrote: "To the young people I have this to say: look around you, you will find subjects for your indignation – the way immigrants, the homeless and the Roma are treated. You will find concrete situations to take you on towards major civic action. Look for them and you will find them!" This message can just as well be addressed to cartographers. Because it takes on the garb of authority, a map is a weapon that can carry weight in public debate. A graph brandished in a television broadcast can put a stop to any well-constructed argument. A map shown on a screen puts a seal on a reality that is then seen as established. At a time when we are dominated by communication and immediacy, the confrontation of ideas is also found in the confrontation of images.

If you set out to make maps, this demiurgic power requires you to enter into this debate. Maps are not just illustrations to make a text more attractive. They are not just "pretty". They shape arguments. They can influence, warn, and give incentive for action. So, on your marks, get set, and go to your computers, your pencils, your tablets, the map war is declared. Create otherness, be controversial, dispute, draw your own view of the world. In other words, produce maps that shift boundaries and wage the cartographic battle.

● FOCUS: Radical Cartography

Far from the world of experts and academics, there is an unusual brand of cartography. Mingling art, science, and social activism, this cartography is the domain of artists, architects, and geographers with a commitment.

Whether radical, critical, contradictory, militant, heterodoxic, or civic, these maps can be of many kinds. But what characterizes them above all is the central notion that citizen needs to appropriate the power of maps. Thus for cartographers, the idea is that the time has come to replace official propaganda by protest, showing the invisible structures of a dominant order and using a map to denounce it. This type of mapmaking aims to undermine the structures of power and domination.

In this counter-cartography movement, the rules of graphic semiology ultimately are not very important. What matters is the artistic expression and the symbolic impact of the map. Representations can be iconoclastic or comply strictly with the rules of graphic semiology. This is not the point. Radical cartography is a free form of expression, where what matters is the impact of the message. It is not just a specific way of drawing a map; this critical approach to the map is seen above all in its finality. The radical map is never an end. It is not decorative. It underpins an argument, for the purpose of concrete action. It rings alarm bells, or triggers political action, action in the field, or makes demands. It is a militant form of mapping serving the downtrodden, a tool for contestation and the regaining of power that has been confiscated.

## THE MAP GAME PART 3: DRESSING AND STAGING A MAP

This is where all the earlier choices can be called into question.
   The map is to be published online on an opinion website.

### 1. Initial choices

• Choose an opinion website
• Identify the editorial approach
• Determine the audience to be targeted.

### 2. Layout

• Assemble all the elements to "dress" the map.
• Choose the layout that will fit the graphic format of the website (colors, font, etc.).
• Refer to the downloaded metadata to draft the source information.
• Pay attention to the hierarchization of the different elements.
• Draft the title in such a way as to comply with the editorial approach and to make users want to look at the map.

### 3. Staging – moving further forward

Stand back and ask yourself a few questions:

• What does the map constructed in this manner have to say?
• Does the basemap (projection, generalization, grid, etc.) meet the bill?
• Is the graphic semiology adopted suitable?
• Are there any superfluous elements on the map (whether in the image or among the different elements used to "dress" it)?
• Could certain elements reinforce the understanding and/or the effectiveness of the message delivered (reading cues, toponyms, lines, etc.)?
• If necessary, return to steps 1 and/or 2.

# General Conclusion

*"Make maps, not photos or drawings"*

Gilles Deleuze and Félix Guattari, 1980

From the Nuzi tablets (2200 B.C.) to Google Maps (2004 A.D.), map designs have evolved dramatically with time. Through scientific and technical progress, the mapmaking adventure has always been characterized by a race towards precision, in order to localize places with ever more detailed methods. Today, as the world is being entirely digitalized in 3D by Google or by the Open Street Map project, we are finally coming to the end of this quest that humanity has been leading for centuries. But does it necessarily mean the end of the cartographic epic? While it is now possible to store a more and more precise digitalized version of the world on computer, it does not necessarily make that digitalized reality more intelligible. This is where the new challenges lie in terms of thematic cartography. Without questioning the obvious need for and usefulness of information technology storage systems, the real cartographic challenge today is to make geographical data intelligent.

The twenty-first century cartography is a discipline at a crossroads with science, technology, ethics, politics, and even art. A hybrid discipline, between spatial statistics, geomatics, computer graphics, etc., cartography is connected to a great number of professional spheres. Used in several domains – geomarketing, territorial planning, and epidemiology, the environment, climatic risks, etc. – cartography is a discipline that is difficult to pin down. Whether a geomatics engineer, a developer, a statistician, a computer graphics designer, or an artist, the cartographer has many modes of existence.

These two approaches to design a map are of course closely linked and overlapping. A scientific map can be "staged" in efficient and artistic manner; a communication map can be constructed with compete scientific thoroughness. Thus, the two approaches are complementary and function alongside, combining exploratory method with means of expression.

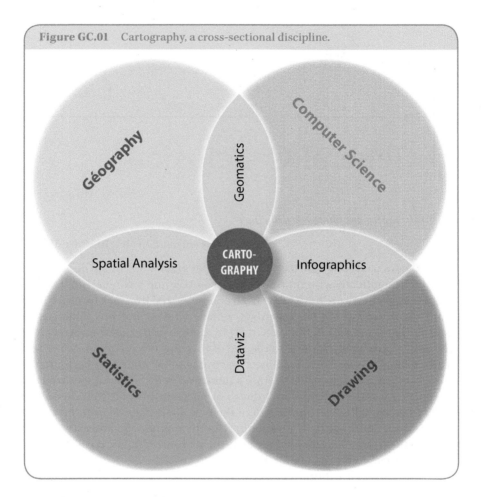

**Figure GC.01**    Cartography, a cross-sectional discipline.

So, what is the real identity of cartography? Cartography is the art of laying out maps. From the data selection, understanding it, processing it, representing it, to laying it out, a lot of knowledge and know-how is involved. Not just anybody can become a cartographer. It takes real professional skills. Theories and concepts have to be learnt and tools have to be mastered before they can be put into practice: spreadsheets, statistical tools, geographical information systems, cartography software programs, computer-assisted drawing, etc.

However, cartography is not only purely technical. Above all, it is the art of formalizing geographical space and of making visible what the eye cannot see. Cartographers handle data, localize it, interpret it, and stage it, as with a play. They depict space according to hypotheses and objectives. Through their eyes, they gaze at the world and let us see it. Unlike a table with figures and numbers, a map thus adds to raw, unprocessed knowledge a perception of the world, ideas, and values. When a map is made, reality is never transcribed faithfully; it is presented as an image, for which filters are chosen. These filters stem from the cartographers' artistic sensitivity, their knowledge, their affects, their convictions, their technical know-how, and their imagination. Whether it is painted on canvass, drawn with a computer mouse, or generated by geographical information systems or dynamically via an online application, any map is the result of the choices made by its authors. In the end, the author's skillfulness and talent resides in his ability to reveal the deep structures of a complex spatial reality and to build an image capable of telling a story.

With this manual, we hope to rehabilitate the term "cartography". Often taken, wrongly, as old-fashioned, in favor of the more contemporary terms "geomatics" or "Geoviz", cartography has nonetheless some advantages. It is a rich and abundant discipline. When we make maps, we become at the same time explorers and pedagogues. Spatial structures are explored, we strive to understand how geographical space works, we analyze it, and we dissect it. Once the analysis has been carried out, results have to be shared. To do this this, cartographers construct images to tell us about territories. They formalize and depict space; they build a world by materializing the imperceptible.

Any map is an invitation. It is an invitation to see, dream, think, and act. So let us dream, imagine, create, and make maps. And proclaim its legitimacy!

**Nicolas Lambert and Christine Zanin**
**48°49′34,824″ N, 2°22′54,771″ E**

# Annexes

## ANNEX 1: THE MAPMAKER'S TOOLBOX

To succeed, a cartographer needs to master the sequence collection, processing, analysis, and representation of geographical data. To do this, several families of software are available. In this annex, we detail the different categories. In each case, we focus on two or three programs that seem to us to be the most relevant. As far as possible, we give precedence-free and open-source software when genuinely used in the professional field.

### A.1.1 Geographical Information System (GIS)

The term "GIS" is ambiguous, since it refers to both GIS software and to a set of coherent geographical data stored in a database. More broadly again, it sometimes refers to the organized entity comprising data, equipment, procedures, and human resources working on geographical information. The main functions of a GIS are as follows: (1) extraction, (2) capture, (3) management and storage, (4) analysis, and (5) representation of geographical data. While they are not very efficient for representing geographical information, GISs are efficient tools for the production of basemaps. They enable the digitization of basemaps, their generalization, and changes in projection. They can generate a geo-referencing system, and they also enable the display and superimposition of different layers of information from different sources and in different formats.

It should be noted that GISs are not cartographic software. Despite constant progress, their representations of geographical information can be unreliable: symbols are sometimes unclassified and non-proportional, and the default colors are often unsuitable.

- **ArcGIS** is GIS software developed by the American company ESRI. It is probably the most efficient proprietary software on the market. ➲ http://www.esrifrance.fr/
- **Quantum GIS** (QGIS) is multi-platform open-source software. It is user-friendly and efficient, and will enable the mapmaker to carry out all the operations relating to cartographic design. ➲ http://www.qgis.org

## A.1.2 Statistical and Spatial Analysis Software

The production of theme maps always requires a phase of processing and/or analysis of the statistical data. While certain operations can be directly performed using other software programs (descriptive statistics, discretization, etc.), other procedures (multivariate classifications, spatial modeling, etc.) require more specialized software (SAS, SPSS, SPAD, R, XLstat, etc.)

- **Excel (and its extension XLstat)** enables most of the procedures that are useful in cartography to be performed. Although accessible to those who are used to Microsoft Office tools, it is nevertheless fairly costly, and the reproducibility of processing procedures is complicated. ➲ https://www.xlstat.com
- **R** is a programming language dedicated to statistical analyses. It is a free and open-source (license GNU GPL) software, available on all platforms (Mac OS, Windows, Linux), and it has dominated the scene in recent years in the human sciences. R provides by default for basic instructions enabling the commonest operations to be performed. New functions, grouped into packages, can easily be added so as to broaden the scope of the language. The strength of R in comparison with software with graphic interfaces is that the automation of procedures is very simple. ➲ https://www.r-project.org
- **GeoDa** is free software enabling the analysis of spatial data (autocorrelation, spatial modeling, etc.). In terms of geo-visualization, the program also offers a few interesting functions, such as the production of choropleth maps, Thiessen polygons, and cartograms according to the Danny Dorling method (see Chapter 7). ➲ https:// geodacenter.github.io

## A.1.3 Cartography software

Cartography software provides tools that are focused on visualization of the geographical information according to the rules of graphic semiology. There are in fact very few theme mapping software programs as such.

- **Magrit** is a free and open-source online application of cartography which provides conventional cartographic methods coupled with innovative techniques (proportional symbols, choropleth map, discontinuity map, smoothed map, gridded map, cartogram, etc.). The application accepts many input formats. It allows exporting the final map in several formats as well. Produced maps are customizable thanks to a large choice of projections, color palettes, fonts, and more.

Magrit is developed and maintained by Matthieu Viry and the members of UMS RIATE. ➲ http://magrit.cnrs.fr

- **Cartography (R)** is a package that enables the production of theme maps under R environment. It offers a whole range of cartographic representations: proportional symbols (circles, squares, and bars), choropleth maps, and qualitative maps (typologies). It also offers modes of representation that are often difficult to implement on other software: flow maps, discontinuity maps, grids, Bertin's dots, etc. ➲ https://cran.r-project.org/web/packages/cartography/index.html
- **Scape Toad** is free software enabling the creation of cartograms using the Gastner and Newman method (see Chapter 7). This software, based on shapefile format, enables the import, distortion, and export of a basemap in vector format. ➲ http://scapetoad.choros.ch (see also ➲ https://go-cart.io)

## A.1.4 Drawing Software (Computer-Aided Design, CAD)

Tools for drawing assisted by computer are an important part of the software toolbox that the cartographer needs to master. These tools for vector-based graphics enable maps to be "finalized" before publication – layout, "dressing", "staging", and editing. For instance, they enable a choice of colors, changes in the thickness of lines, the positioning of different elements on the map, the hierarchization of the information, or the drafting of text. There are many programs available: Adobe, Illustrator, Freehand, CorelDraw, Xara X, or Inkscape.

- **Adobe Illustrator** is a CAD software program developed by the Adobe Company. It is the most efficient vector-based graphics software, well before the open-source solutions. ➲ http://www.adobe.com/fr/products/illustrator.html
- **Inkscape** is another open-source CAD software based on SVG format. Although Inkscape aims to become the reference open-source CAD tool, it still falls well short of its direct competitor, Adobe Illustrator. Even so, Inkscape is software that is easy to access and enables most of the operations required to produce a map. ➲ https://inlkscape.org

## A.1.5 Online useful tools

When producing thematic maps, the cartographer can also use online services. These concern the generalization of basemaps, geo-coding, the production of colors palettes, and the online posting of interactive maps.

- **Mapshaper** is an online application enabling the interactive simplification of basemaps. The use of this very handy tool does not however preclude the need for human choice for this type of operation (conceptual layout, harmonization, etc.). ➲ http://www.mapshaper.org
- The Census Geocoder is tool available online through an API (Application Programming Interface) that converts address to geographic coordinates (latitude/longitude) all over the US territory. https://www.census.gov/data/developers/data-sets/Geocoding-services.html
- **ColorBrewer** is an online application offering various color palettes free of charge specially intended for mapping (single and dual gradations, typologies). On this website, each palette is assessed according to different criteria – whether they can be perceived by color-blind individuals, visibility once printed, use on LCD screen, etc. ➲ http://colorbrewer2.org
- **"Colored gradients"** is a tool developed by the French geographer Laurent Jégou. It enables the creation of palettes of colors for cartography. The tool requires the choice of a color at the outset, a color to end with, an inflection color, the number of classes required, and the interpolation method (linear or curved). Thus, the tool enables the development of tailored palettes closely suited to the statistical data to be represented. ➲ http://www.geotests.net/couleurs/gradients_inflex.html
- **D3.js** is a JavaScript library enabling the display of static, interactive, and animated vector graphics online. While there are numerous similar libraries, this one has the advantage of offering a certain number of functions related to the dynamic display of geographical information (cartograms, projections, etc.) ➲ http://d3js.org

## A.1.6 Processing Sequences

A map is constructed in several steps. Data acquisition, capture, digitization, processing, statistical analysis, interpretation, representation, layout, etc. To complete these different phases in cartographic design, the mapmaker will often need to switch from one type of software to another to make use of different complementary functions. This each cartographic production can be traced by a sequence of computer-based operations that may or may not enable automation. In short, there are a thousand and one ways to conceive a map, and a multitude of tools to achieve it.

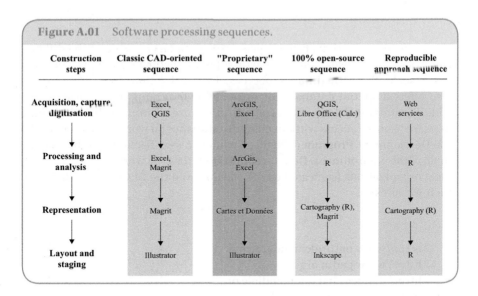

**Figure A.01**   Software processing sequences.

| Construction steps | Classic CAD-oriented sequence | "Proprietary" sequence | 100% open-source sequence | Reproducible approach sequence |
|---|---|---|---|---|
| Acquisition, capture, digitisation | Excel, QGIS | ArcGIS, Excel | QGIS, Libre Office (Calc) | Web services |
| Processing and analysis | Excel, Magrit | ArcGis, Excel | R | R |
| Representation | Magrit | Cartes et Données | Cartography (R), Magrit | Cartography (R) |
| Layout and staging | Illustrator | Illustrator | Inkscape | R |

## ANNEX 2: THE MAPMAKER'S DATA

### A.2.1  Europe

In Europe, EuroGeographics is the main supplier of basemaps. This association, which at present gathers 60 organizations in 46 countries, works on the development of the European infrastructure of interoperable spatial data. UMS RIATE also provides the GREAT basemap free of charge (a generalized basemap of European regions for territorial development). This generalized basemap is directly suited to theme mapping. It takes account of changes in subdivisions across European regions from 1980 to the present. Concerning statistical data, the main supplier of data for Europe is Eurostat. This is the body in charge of statistical information at European level. The European environmental agency also publishes a lot of data on environmental and land-use themes (Corine Land Cover).

### References

⮞ http://www.eurogeographics.org
⮞ http://www.ums-riate.fr/Webriate/?page_id=153
⮞ http://ec.europa.eu/eurostat
⮞ http://www.eea.europa.eu/data-and-maps

## A.2.2 At World Level

Numerous organizations provide geometric data on the scale of the different countries. The website Natural Earth supplies vector and matrix data on a global scale. There are different degrees of generalization. The GADM website also offers a spatial database for infra-national administrative grids world-wide. For statistical data, several organizations are prominent. The World Bank gives free-source access without charge to statistics on world development. The UN Development Program (UNDP) also offers a user-friendly database on the topic of the environment. Data on countries across the world is easy to access. Most organizations (such as the FAO, the IMF, and the OECD.) publish data on their websites.

### References

⮫ http://www.naturalearthdata.com
⮫ http://www.gadm.org
⮫ http://donnees.banquemondiale.org
⮫ http://geodata.grid.unep.ch

## A.2.3 Data Formats

The most widely used data format for describing geographical entities (base-maps) is the shapefile format. This format originated in the area of GIS. It was initially developed by the company ESRI, and rapidly became standard. It is used in most software, whether open source or proprietary. There are nevertheless other formats for geospatial data that are more open (geoJSON, Rdata, etc.), but they are less widely used. In all events, software like QGIS will easily enable conversion from one format to another. The data formats for attribute data take the form of tables. These can be proprietary formats (xls) or open source (csv).

## ANNEX 3: CARTOGRAPHIC TERMS

The definitions proposed in this glossary may differ slightly from those advanced throughout this handbook. We have favored here those commonly found in generalist books published. They provide additional insight into the concepts discussed. They come from several sources, French and American: French Committee of Cartography (CFC), International Cartographic Association, Roger Brunet (1990 and 1992), Arthur H. Robinson (1995), Christine Zanin and Marie-Laure Trémélo (2003), Judith A. Tyner (2010), Menno-Jan Kraak and Ferjan Ormeling [2010] Michèle Béguin and Denise Pumain (2014), and Kenneth Field (2018).

**Abacus (Lenz-Cesar):** A figure enabling the determination of the surface area of simple shapes (squares, circles) to design proportional maps. But since the use of computer graphics in mapping, the abacus has been little used.

**Additive primaries:** They are the three basic colors (red, green, and blue), often shortened as RGB. When mixed (added) together in equal amounts at 100% intensity, white light is created. When added together in different proportions, all other colors can be created.

**Anamorphosis:** See cartogram.

**Areal phenomena:** Geographical phenomena that extend over defined areas.

**Associative symbol:** A type of symbol that expresses association and differentiation between phenomenon.

**Atlas:** A book of maps or charts (Oxford English Dictionary); an ordered collection of maps designed to represent a given space and illustrate one or several themes (French Committee of Cartography).

**Attribute data:** Data is associated with an object or a place, to describe object and localize information. It corresponds to the semantic level of geographical information.

**Azimuthal projection:** Projection that shows azimuth correctly from the center point; great circles through the center point are straight lines.

**Basemap:** A map having only essential outlines and used for presentation of specialized data of various kinds. In thematic mapping, the basemap is the "container" for the geographical information to be represented. A map used as a primary source for compilation (Source map) or a framework on which new detail is designed. A good basemap enhances the information you are showing and the user experience of your map.

**CAC:** Computer Assisted Cartography

**CAD:** Computer-aided design (or drawing)

**CAM:** Computer-aided mapping

**Cartogram (or anamorphosis):** A cartographic representation where the spatial units (geographical entities or grid divisions) are distorted according to a quantitative data. There are several types of cartogram. An abstracted and simplified map for which the base is not true to geographic scale.

**Cartography (mapmaking, mapping):** The art, science, and technology of making maps. This refers to the studies and the scientific, artistic, and technical operations that are performed on the results of direct observation or using documentation.

**Choropleth map:** A geographical representation that uses areas of color or different shades of gray. Quantitative map on which statistical or administrative areas are colored or shaded proportionally to the relative value represented.

**Clarity:** It refers to ease of understanding the maps, and all elements on the map serve a designed purpose.

**Classification:** Placing data into groups that have similar characteristics or values. Classification is also part of a generalization process.

**Clipping:** An operation consisting in removing the contours of objects, a procedure often used in cartography so that superimposed proportional symbols can be seen.

**CMYK:** The subtractive primary colors – cyan, magenta, and yellow, plus black.

**Color:** The quality of an object with respect to light reflected by the object, usually determined visually by measurement of hue, saturation, and brightness of the reflected light; saturation or chroma; hue. The visual variable "color" refers to the way in which hue, saturation, and/or intensity of a figuration is made to vary.

**Composition:** The arrangements within the borders of map elements, such as subject area, title, legend, and scale.

**Compromise projection:** A projection that has no special properties.

**Conformal projection:** Projection on which the shapes of very small areas are preserved. With the conformal projection, parallels and meridians cross at right angles, and scale is the same about a point.

**Conic projection:** Projection that appears to have been projected into a cone.

**Contiguous cartogram:** A value-by-area cartogram in which the different areas represented preserve boundaries.

**Contour:** The contour is what describes the outline and shape of the object. In computer graphics, an object is defined by its contour and its ground.

**Contrast:** Differences in light and dark shades, thick and thin lines, large or small types, rough or smooth textures, or patterns on a map.

**Conventional projection:** Any type of projection except azimuthal, cylindrical, and conic projections. Also called "mathematical" projections.

**Cylindrical projection:** Projection that has been created by projecting the earth's grid into a cylinder.

**Cover:** It defines the spatial limits of a cartographic representation. Also called "spatial extent".

**Data classification:** Grouping of data into categories or numerical classes.

**Datum:** Any value that serves as a reference or base. For contour maps, the datum is sea level.

**Design:** The process of creating maps. The appearance of a map.

**Digitization:** Digitization is a stage in computerized graphic processing where each place can be identified by geo-referenced coordinates. A basemap in digital form derived from a raster image (aerial photo, satellite image, scanned image, etc.).

**Discontinuity map:** A map showing spatial breaks across a network of subdivisions.

**Discontinuous distribution:** A distribution that does not occur everywhere in the mapped area.

**Diverging color scheme:** A color scheme used to represent increases and decreases of a variable from a midpoint, such as positive and negative changes.

**Discretization:** This procedure consists in dividing up a statistical series into classes using a specific method – quantiles, nested means, equal spans, amplitudes, etc.

**Dorling cartogram:** A value-by-area cartogram on which the enumeration areas are replaced by uniform geometric figures, usually circles, which are drawn proportional to the value represented.

**Dot density map:** A variation of the dot map that places dots randomly within the enumeration area.

**Dot map:** A representation of geographic phenomena on which dots represent a specified number of the phenomenon being mapped.

**Dressing:** For a geographical image to become a complete map, it needs to be "dressed" with different outside elements to assist the reading of the map (title, legend, scale, etc.).

**Earth's grid:** The system of parallels and meridians on earth or a globe. Also called the "graticule".

**Equal-area projection:** A map projection that preserves a uniform area scale. Also called an "equivalent" projection.

**Equator:** An imaginary great circle drawn around the surface of the earth midway between the north, and south poles. It divides the earth into two equal hemispheres. It's designed as 0° of latitude.

**Equidistant projection:** Projection that shows distance correctly along certain lines, or from certain points.

**Equivalent projection:** Projection on which countries, or any areas maintain their correct area scale. Also called "equal-area" projection.

**Ellipsoid:** The mathematical surface area enabling the modeling of terrestrial space.

**Exaggeration:** An operation of generalization in which features are made larger.

**Extrusion:** A process consisting in creating a three-dimensional figuration from a two-dimensional representation.

**Figuration:** A graphic construction made up of simple elements (points, lines, zones).

**Format:** The size and shape of a map.

**Geographic data:** Facts that measure or describe aspects of geographical phenomena.

**Geographic information system:** A computer-based system for collecting, managing, analyzing, modeling, and presenting geographic data for a range of applications.

**Geoid:** An earth-shaped figure. The figure of the earth viewed as a mean sea-level surface.

**Geometric symbol:** An abstract symbol that uses geometric shapes such as circle, triangles, squares, and cubes.

**Generalization:** A process of simplification based on selection criteria in terms of number and quality when reducing a scale. A reasoned reduction of the representation of elements on the map. Systematic simplification of contours. Any map, whatever its scale, involves a generalization of the representation. It is possible to distinguish the generalization of lines by simplification (required when the scale is reduced) or generalization via classification (discretization) of statistical data.

**Geographical image:** An image of a geographical phenomenon or feature represented on a basemap. It is a map without elements to "dress" it.

**Geographical information:** Information on places, or information that is localized. By extension, it is spatialized data relating to a set of places.

**Geographical object:** Something that has a dimension in space, involves places, and is studied in geography (Brunet). Also called "geographical unit".

**Geo-referencing:** The process of taking a digital image and adding geographic information to the image so that GIS or mapping software can "place" the image in its appropriate real-world location. Geometrical adjustment procedure consisting in making chosen points correspond to the same points on a different document (overlaying).

**Grain:** It refers to the enlargement or reduction of a texture-structure. This means varying the size of the elements making up a ground, with a constant ratio of black to white.

**Graphic charter:** A document describing the rules applied for designing graphic constructions (layers, color palette, dressing, line thickness, etc.).

**Graphical scale:** A graduated line marked in ground units that allows distances to be measured from a map. Also called a "bar scale".

**Graticule:** A network of coordinates resulting from a projection and forming a squared base. Grids can also be regular quadrilaterals. Grid of parallels and meridians on the earth.

**Ground:** A construction made up of simple, separate, graphic elements repeated across a whole surface area in regular and identical manner. As in "background".

**Halo:** A highlighting effect around a graphic element in order to bring it to prominence.

**Harmonization:** This is a general operation of homogenization of the basemap to obtain a comparable level of generalization at any point on the map.

**Hemisphere:** The two halves of the globe on either side of the equator.

**Histogram:** A graph with bars in which the area of the bars is proportional to the frequency in each class.

**Hue:** The dimension of color related to its wavelength. Main characteristic of a color which distinguishes it from other colors

**Inset map:** A map to complement the main map. It is generally an extract from the main map, or a localization map.

**Intensity:** The richness of a color. Also called "saturation".

**Isochrone:** A line that joins all points having the same time from a specific point.

**Isoline:** A line that joins all points having the same value between two categories.

**Label:** Name of a geographic feature.

**Latitude:** The angular distance of a point on the Earth from the Equator. The angle is measured in degrees. The North Pole has a latitude of 90°N. All points with the same latitude are on the same parallel.

**Layer:** It is a structuring of data enabling the superimposition of geographical objects on a map. In geomatics, a layer refers to a group of point, line, or zone objects that have a relationship one with the other.

**Layout:** The way in which the different elements used to dress a map are positioned on the page.

**Legend:** A key or list of explanations for the various figurations on a map.

**Lettering:** Map lettering is the task of selecting fonts and font styles placing (or positioning) labels and text on a map.

**Lightness:** Shades of a hue from light to dark. Also called "value", a visual variable.

**Linear cartogram:** Cartogram that is concerned with distance.

**Linear symbol:** A line used to represent geographical phenomena that are linear in nature.

**Longitude:** The longitude of a place is the angle measured in degrees formed between the meridian passing through this place and the Greenwich meridian. There are 180°W and 180°E of longitude in relation to the Greenwich meridian.

**Map:** A diagrammatic two-dimensional representation of an area of land or sea; a conventional geometric representation, generally flat, showing relative positions of abstract or concrete features, localizable in space. A map is made up of a geographical image and different elements to dress it.

**Metadata:** Data documenting a dataset: data sources, date of capture, geodesic referencing, cartographic projection, digitization method and scale, data structure, precision, file format, etc.

**Meridian:** A line that connects all points having the same longitude. Meridians converge at the poles.

**Monochrome:** Shades on a given color (for instance, from light to dark blue). The combination of two visual variables, color, and value.

**Noise:** Anything that interferes with the map communication.

**Nominal measurement:** Classification based on the nature of the phenomena without any indication of quantity.

**Noncontiguous cartogram:** A type of value-by-area cartogram on which the areas represented are separate from one another. Internal boundaries are not retained.

**Opaqueness:** A value that describes how far an object enables another object that is overlaid by it to be perceived. An object with an opaqueness value of 1% is transparent and almost invisible, and object with a 100% opaqueness value is completely opaque.

**Order:** Arrangement of graphic elements in a logical manner.

**Ordinal measurement:** Classifies and ranks data without specifying numerical values.

**Orientation:** (1) Angular indication for the top of the map. Generally indicated by the north (if possible parallel to the frame of the map). In Medieval cartography in the Western world maps were literally "oriented" (to the east) – the top of the map indicated the direction of the Holy Land. (2) The visual variable "orientation" refers to the angle between a figuration and the vertical.

**Palette:** It refers to a coherent, organized set of colors or shades of gray.

**Parallel:** A line that joins all points having the same latitude. Parallels are true east–west line that encircles the globe.

**Perspective projection:** A projection that can be developed geometrically from a generating globe.

**Pictorial symbol:** A symbol on a map that is a small picture of some object. Also called "mimetic symbol".

**Pie chart:** A circular symbol divided into sectors to indicate proportions of a total value. Also called "pie graph" or "sectored circle".

**Plane projection:** A transformation of the earth's grid onto a plane surface.

**Point phenomena:** Geographic phenomena that occur at discrete points.

**Primaries:** The three colors used to create all other colors. There are additive primaries (red, green, and blue) and subtractive primaries (cyan, magenta, yellow, and black).

**Projection:** A systematic arrangement of all or a part of the earth's grid on a plane.

**Proportional symbol:** A point symbol, such as a circle or square that is drawn so that its area is visually proportional to the amount represented.

**Qualitative symbol:** Symbol that shows some nonquantitative aspect of a geographic phenomenon.

**Ratio measurement:** Classifies and gives references between values using a scale that starts at absolute zero.

**RGB:** The additive primary colors (red, green, and blue).

**Saturation:** The colorfulness of a color.

**Scale:** The scale of a map expresses the reduction between real distances and their representation on paper. A scale of 1:100 means that the drawing is 1/100th of the size of the real thing. A scale of 1:10,000 means that 1 cm on the map represents 10,000 cm (100 m) on the ground. A large-scale map represents a small area.

**Selection:** A generalization process that involves choosing the information that will be shown in the map.

**Semiotics:** The theory of signs and symbols.

**Shading or shadow:** A highlighting effect or a graphic procedure used to make an object or element on the map stand out.

**Shape:** The visual variable "shape" refers to variations in the contours of a figuration without change in proportions.

**Simplification:** A generalization process that involves elimination of unnecessary detail.

**Size:** The visual variable "size" refers to variations in the length, surface area, or volume of a figuration.

**Smoothing:** A smoothed map is one where space is represented in continuous manner by way of isolines or isosurfaces. It is generated using different methods of spatial interpolation.

**Spatial data:** Information for geographic or spatial phenomena.

**Spatial subdivisions:** Subdivisions of space without overlaps. The subdivision ranges from plots or fields to states, across all geographical scales.

**Staging:** The positioning of different elements making up a map so as to produce a narrative. The reference is her obviously theatrical.

**Supplemental text:** A small frame placed beside the main map, giving the names of the authors, technical elements relating to the design and production, the publisher's name, the logo, the date, and/or any other information that may be required.

**Symbol:** A mark placed on a map that by convention, usage, or reference to a legend is understood to represent a specific feature.

**Symbolization:** The process of designing or selecting symbols for a map.

**Texture-structure:** The visual variable "texture-structure" refers to the combination of graphic elements covering a surface and forming a ground.

**Thematic map:** A map that features a single distribution, concept, or relationship for which the base data serve only as a framework to locate the distribution being mapped.

**Topology:** The study of surfaces. Mathematical study of the properties that are preserved through deformations, twisting, and stretching of objects.

For example, a circle is topologically equivalent to an ellipse (into which it can be deformed by stretching), and a sphere is equivalent to an ellipsoid. It is a branch of mathematics derived from the study of non-metric properties, including the notions of boundary and neighborhood (contiguity).

**Toponymy:** The science of place names. On a map, the toponyms are the set of place names used, and the toponym is the actual name of a place.

**Trapping:** This is applied to a graphic element when it needs to stand out from another when they overlap or coincide. In cartography, it is used to make text elements or symbols more prominent when they cut across or overlay a line.

**Value:** A quantity to be represented. The visual variable "value" refers to variation in the proportions of black and white on a given surface area.

**Vector:** Digitized data in the form of geometric figures. A vector image is made up of arcs, nodes, and segments.

**Vectorization:** The operation consisting in converting a raster image into a vector image.

**Visible spectrum:** That part of the electromagnetic spectrum that is visible to the human eye (from violet at the short-wavelength end through indigo, blue, green, yellow, and orange, to red at the long-wavelength end).

**Visual hierarchy:** It refers to different levels of emphasis of the element of a graphic design.

**Visual variable:** A graphic device enabling the data represented to be differentiated on the map. These variables concern shape, dimension, value, orientation, or color. Each visual variable has its own characteristics and will be suitable for the representation of a given spatial distribution.

**Visualization:** Any technique for creating images, diagrams, or animations to communicate a message. The representation of an object, situation, or set of information as a chart or other image. Exploration of data and seeing it in different ways.

# Bibliography

Andrienko, Gennady, and Andrienko, Natalia. (2006). *Exploratory Analysis of Spatial and Temporal Data*. Berlin: Heidelberg, Springer-Verlag.

Béguin, Michèle, and Pumain, Denise. (2014). *La représentation des données géographiques. Statistique et Cartographie* (5th ed.). Paris: A. Colin.

Bertin, Jacques. (1967). *Sémiologie Graphique. Les diagrammes, les réseaux, les cartes*. Paris: La Haye, Mouton, Gauthier-Villars. (1973) 2nd ed.; (1999) 3rd ed. 2e édition: 1973, 3e Paris: EHESS.

Bertin, Jacques. (1983). *Semiology of Graphics* (William J. Berg, Trans.). Madison, WI: University of Wisconsin Press.

Borges, Jose-Luis. (1982). *L'auteur et autres textes* (3rd ed.). Paris: Gallimard.

Brewer, Cynthia A. (1994). Color use guidelines for mapping and visualization. In Allan M. MacEachren and D.R. Fraser Taylor (Eds.), *Visualization in Modern Cartography* (pp. 123–147). New York: Pergamon.

Brewer, Cynthia A. (2005). *Designing Better Maps: A Guide for GIS Users*. Redlands, CA: ESRI Press.

Brotton, Jerry. (2014, reprint). *A History of the World in Twelve Maps*. London, UK: Penguin Books.

Brunet, Roger. (1990). *La carte, mode d'emploi*. Paris: Reclus-Fayard.

Brunet, Roger, Ferras, Robert, and Théry Henri. (1992). *Les mots de la géographie, dictionnaire critique*. Paris: Reclus, La Documentation française.

Campbell, John. (2001). *Map Use and Analysis* (4th ed.). New York: McGraw-Hill.

Cartwright, William, Milelr, Suzette, and Pettit, Christopher. (2004). Geographical visualization: Past, present and future development. *Spatial Science*, 49, 5–36.

Cartwright, William, Peterson, Michael P., and Gartner, Georg (Eds.). (2007). *Multimedia Cartography* (2nd ed.). New York: Springer-Verlag.

Cauvin, Colette, Escobar, Francisco, and Serradj, Aziz. (2008). *Cartographie Thématique*. Vol. 1–5. Paris: Information géographique et Aménagement du territoire, Hermes, Lavoisier.

Cauvin, Colette, Escobar, Francisco, and Serradj, Aziz. (2010a). *Cartography and the Impact of the Quantitative Revolution*. Vol. 2, London, UK: Iste Ltd, Wiley & sons, Inc.

Cauvin, Colette, Escobar, Francisco, and Serradj, Aziz. (2010b). *New Approaches in Thematic Cartography*. Vol. 3. London, UK: Iste Ltd, Wiley & sons, Inc.

Churchill, Winston. (1946). The sinews of peace. *Iron Curtain Speech*, 1946, March 5th. Fulton. https://www.youtube.com/watch?v=X2FM3_h33Tg.

Claval, Paul. (2011). Histoire de la Géographie. *Que sais-je?* n°65. https://www.cairn.info/histoire-de-la-geographie--9782130586524.htm

Cleveland, William S. (1994). *The Elements of Graphing Data*. Murray Hill, NJ: ATT Bell Laboratories.

Crampton, Jeremy W. (2001). Maps as social constructions: Power, communication and visualization. *Progress in Human Geography*, 25(2), 235–252.

Crampton, Jeremy W., and Krygier, John. (2006). An introduction to critical cartography. *ACME: An International E-Journal of Critical Geographies*, 4(1), 11–33.

Dent, Borden D., Torguson, Jeffrey, and Holder, Thomas W. (2009). *Cartography: Thematic Map Design* (6th ed.). New York: McGraw-Hill Higher Education.

DiBiase, David. (1990). Visualization in the earth science. *Earth and Mineral Science Bulletin*, 59(2), 13–18.

Dodge, Martin, McDerby, Mary, and Turner, Martin. (Eds.). (2008). *Geographic Visualization: Concepts, Tools, and Applications*. Hoboken, NJ: Wiley.

Dorling, Danier, and Fairbairn, David. (1997). *Mapping: Ways of Representing the World*. New York: Prentice Hall.

Dorling, Denis. (1993). Map design for census mapping. *The Cartographic Journal*, 30(2), 167–183.

Dorling, Denis. (1994). Cartograms for visualizing human geography. In D.J. Unwin and H.M. Hearnshaw (Eds.), *Visualization in Geographic Information Systems* (pp. 85–102). London, UK: Wiley & sons, Inc.

Edney, Matthew, H. (2005). *The Origins and Development of J. B. Harley's Cartographic Theories, Cartographica Monograph* 54, *Cartographica* 40, nos. 1 & 2. Toronto: University of Toronto Press.

Field, Kenneth. (2018). *Cartography*. New York: ESRI Ed.

Goodman, Nelson. (1972). *Problems and Projects*. Indianapolis: Bobbs-Merrill.

Gould, Peter, and Bailly, Antoine. (1995). *Le pouvoir des cartes – Brian Harley et la cartographie*. Paris: Economica.

Grasland, Claude, and Madelin, Malika. (Eds.). (2007). *The Modifiable Area Unit Problem, Final Report of Espon Project 3.4.1.* Luxembourg: The ESPON Monitoring Committee.

Goodchild, Michael F. (1997). A GIS perspective. In *Proceedings, Atlantic Institute Think Tank V: Global Educational Paradigms in Geomatics/Geoinformation* (pp. 77–82). Paris: Ecole Nationale des Sciences Geographiques.

Goodchild, Michael F. (2008). Statistical perspectives on geographic information science. *Geographical Analysis*, 40(3), 310–325.

Harley, J. Brian. (1989). Deconstructing the map. *Cartographica*, 26(2), 1–20.

Harley, J. Brian. (2001). *The New Nature of Maps: Essay in the History of Cartography* (Paul Laxton, Ed.; Introduction by J.H. Andrews). Baltimore, MD: Johns Hopkins University Press.

Harley, J. Brian, and Woodward, David. (Eds.). (1987). *The History of Cartography*. Chicago, IL: The University of Chicago Press.

Harrower, Mark. (2003). Tips for designing effective animated maps. *Cartographic Perspectives*, 44, 63–65, 82–83.

Harrower, Mark. (2009). Cartographic Animation. In Rob Kitchin and Nigel Thrift (Eds.), *International Encyclopedia of Human Geography*. Oxford, UK: Elsevier.

Harvey, Francis. (2008). *A Primer of GIS: Fundamental Geographic and Cartographic Concepts*. New York: Guilford Press.

International Cartographic Association. (1973). *Multilingual Dictionary of Technical Terms in Cartography*. Wiesbaden: Franz Steiner.

Jacob, Christian. (1992). *L'empire des cartes: Approche théorique de la cartographie à travers l'histoire*. Paris: Albin Michel.

Kaiser, Ward L., and Wood, Denis. (2001). *Seeing Through Maps: The Power of Images to Shape Our World View*. Amherst, MA: ODT, Inc.

Korzybski, Alfred. (1933). *Science and Sanity. An Introduction to Non-Aristotelian Systems and General Semantics* (pp. 747–761). Lancaster, PA: The International Non-Aristotelian Library Pub. Co.

Kraak, Menno-Jan. (2007). Cartography and the Use of Animation. In William M. Cartwright, Michael Paterson, and Georg Gartner (Eds.), *Multimedia Cartography* (pp. 317–326). New York: Spinger-Verlag.

Kraak, Menno-Jan, and Brown, Allan. (Eds.). (2001). *Web Cartography: Developments and Prospects*. New York: Taylor & Francis.

Kraak, Menno-Jan, and Ormelling, Ferjan. (2010). *Cartography: Visualization of Geospatial Data* (4th ed.). New York: Prentice Hall.

Krygier, John, and Wood, Denis. (2005). *Making Maps: A Visual Guide to Map Design for GIS*. New York: Guilford Press.

Lacoste, Yves. (1976). *La géographie, ça sert, d'abord, à faire la guerre*. Paris: Maspero.

Lambert, Nicolas and Zanin, Christine. (2016) *Manuel de cartographie – Principes, méthodes, applications (french)*. Paris: Armand Colin.

Lambert, Nicolas and Zanin, Christine. (2019) *Mad Maps: L'atlas qui va changer votre vision du Monde (french)*. Paris: Armand Colin

Linford, Chris. (2004). *The Complete Guide to Digital Color: Creative Use of Color in the Digital Arts*. New York: Harper Collins.

Mac Eachren, Alan M. (1991). The role of maps in spatial knowledge acquisition. *The Cartographic Journal*, 28, 152–162.

Mac Eachren, Alan M. (1994a). *Some True with Maps: A Primer on Symbolization and Design*. Washington, DC: Association of American Geographers.

Mac Eachren, Alan M. (1994b). Visualization in modern cartography: Setting the agenda. In Alan M. MacEachren and D.R. Fraser Taylor (Eds.), *Visualization in Modern Cartography* (pp. 1–12). Oxford, UK: Elsevier.

Mac Eachren, Alan M. (1995). *How Maps Works: Representation, Visualization, and Design*. New York: Guilford Press.

Mac Eachren, Alan M., and Taylor, D. R. Fraser (Eds.). (1994). *Visualization in Modern Cartography*. New York: Pergamon.

Mathian, Hélène, and Sanders Lena. (2014). *Spatio-temporal Approaches: Geographic Objects and Change Process*. Paris: ISTE Ltd.

Monmonier, Mark. (1977). *Maps, Distortions, and Meaning*. Washington, DC: Association of American Geographers.

Monmonier, Mark. (1996). *How to Lie with Maps* (2nd ed.). Chicago, IL: University of Chicago Press.

Monmonier, Mark. (2004). *Rhumb Lines and Map Wars: A social History of the Mercator of Projection*. Chicago, IL: University of Chicago Press.

Muehrcke, Philip C., and Muehrck, Juliana O. (1992). *Map Use: Reading, Analysis, Interpretation* (3rd ed.). Madison, WI: JP Publications.

Openshaw, Stan, and Taylor, Peter, J. (1981). The Modifiable Areal Unit Problem. In N. Wrigley and R.J. Bennett (Eds.), *Quantitative Geography: A British View*. London, UK: Routledge & Kogan Page.

Peuquet, Donna J. (2002). *Representations of Space and Time*. New York: Guilford Press.

Peterson, Michael P. (2003). *Maps and the Internet*. New York: Elsevier.

Reclus, Elisée. (1903). On Spherical Maps and Reliefs. *The Geographical Journal*, 22(3), 290–293.

Rekacewicz, Philippe. (2014). Entre imaginaire et réalité: l'intention cartographique. *Revue 303* n°133. https://www.editions303.com/le-catalogue/numero-133-cartes-et-cartographie/

Robinson, Arthur H. (1952). *The Look of Maps: An Examination of Cartographic Design.* Madison: University of Wisconsin Press.

Smith, Adam. (1776). An Inquiry into the Nature and Causes of the Wealth of Nations. Ed. W. Strahan and T. Cadell, London

Stewart, John Q. (1942). Measure of the influence of a population at a distance. *Sociometry*, 5(1), 63–71.

Tobler, Waldo. (1973). Choropleth Maps without Class Intervals. *Geographical Analysis*, 5, 262–265.

Tobler, Waldo. (2004). Thirty-Five Years of Computer Cartograms. *Annals of the Association of American Geographers*, 941(1), 58–73.

Tyner, Judith A. (1992). *Introduction to Thematic Cartography.* Englewood Cliffs, NJ: Prentice Hall.

Tyner, Judith A. (2005). Elements of Cartography: Tracing 50 Years of Academic Cartography. *Cartographic Perspectives*, 51, 4–13.

Tyner, Judith A. (2010). *Principles of Map Design.* New York: Guilford Publications.

Wood, Denis, and Fels, John. (1992). *The Power of Maps.* New York: Guilford Press.

Zanin, Christine, and Tremelo, Marie-Laure. (2003). *Savoir-faire une carte: Aide à la conception et à la réalisation d'une carte thématique univariée.* Paris: Belin.

Zelevansky, Lynn. (2008). *An Atlas of Radical Cartography.* Los Angeles, CA: Journal of Aesthetics and Protest Press.

# Index